God Made EVERYTHING

Science/Worldview | Level 1

Author: Tamela Sechrist

Editors: Kevin Swanson, R.A. Sheats, Kayla White

God Made EVERYTHING

Hi! I'm Beamer.
I'm a light beam and
I like to shine!

It's going to be great to
learn about God's creation
with you in this book.

Printed in the United States of America.

ISBN: 978-1-954745-18-6

Cover Design: Justin Turley
Interior Design: Justin Turley and Sarah Lee Bryant

Published by:
Generations
19039 Plaza Drive Ste 210
Parker, Colorado 80134
Generations.org

For more information on this and other titles from Generations,
visit Generations.org or call (888) 389-9080.

TABLE OF CONTENTS

INTRODUCTION

This introductory science course for young school-age children is designed to bring to light the love, wisdom, and power of God that is evident in His creation. *God Made Everything* presents the amazing way each created thing works perfectly and how all the created things work together harmoniously in a way that only God could have brought forth.

Beamer, a friendly light beam who loves God's light, appears often in the course, both to keep the child's interest and to illuminate principles and interesting facts.

God Made Everything is designed to work with the way God made children to learn. Children learn best by using a variety of their senses and through hands-on activity. This curriculum helps children learn through:

- **Listening:** The textbook is designed to be read aloud to the child. The language is simple and easy to understand. Important words are in bold so the reader knows to emphasize them and maybe take a detour to a definition box or picture.
- **Seeing:** The beautiful pictures, fun illustrations, and simple diagrams in the textbook provide a visual aspect to learning.
- **Moving:** The *God Made Everything* companion activity book is full of hands-on learning that involves the senses and muscles with a variety of fun activities that are not too burdensome for the parent/teacher. The science topics are reinforced with a balance of action, observation, experiments,

imagination, logic, Scripture, art, cooking, poetry, math, stories, exercise, music, and a little bit of writing. Each activity reinforces the material introduced in the textbook, and every exercise is numbered for easy reference.

How To Use *God Made Everything*

God Made Everything is divided into nine units of four chapters each. Each unit has a memory verse and a children's hymn that children will have an opportunity to work on in the activity book.

The *God Made Everything* books are organized in a way to enable children to internalize what they have learned. Learning comes easiest in small doses with time between each session. Children solidify what they have learned as they play. It becomes permanent as they sleep. To enable this, the following schedule is suggested:

One chapter per week for 36 weeks. Each week:

- **Day 1:** Read aloud the first section of the textbook chapter. Complete the corresponding activity in the activity book (as announced at the end of that section of text).
- **Day 2:** Break
- **Day 3:** Read aloud the second section of the textbook chapter. Complete the corresponding activity in the activity book.
- **Day 4:** Break
- **Day 5:** Read aloud the third section of the textbook chapter. Complete the corresponding activity in the activity book.

May God be glorified, and may you be richly blessed as you study His creation in *God Made Everything*!

UNIT 1

Light

I like being a light beam! I like to chase darkness away. I play with the raindrops to make beautiful rainbows. I can bounce off your toys and into your eyes so you can see to play. I like being light most of all because God talks about it a lot.

Here is a verse to memorize and a hymn to sing as we learn about light!

Memory Verse

Arise, shine;
For your light has come!
And the glory of the LORD
is risen upon you.

(Isaiah 60:1)

Hymn Singing

God, Who Made the Earth

God, who made the earth,
The air, the sky, the sea,
Who gave the light its birth,
Careth for me.

God, who made the grass,
The flower, the fruit, the tree,
The day and night to pass,
Careth for me.

God, who made the sun,
The moon, the stars, is He
Who, when life's clouds come on,
Careth for me.

God, who sent His Son
To die on Calvary,
He, if I lean on Him,
Will care for me.

You can listen to this hymn by searching for "God Who Made the Earth children's hymn" on the internet.

CHAPTER 1

What Is Light?

Light is an amazing, important invention of God! God made **light** on the first day of creation.

In the beginning God created the heavens and the earth. The earth was without form, and void; and darkness was on the face of the deep. And the Spirit of God was hovering over the face of the waters. Then God said, "Let there be light"; and there was light. (Genesis 1:1-3)

As light, Beamer has important jobs to do. He likes doing them and he does them well.

Beamer's List of Light's Jobs

- To give sight to people and animals
- To keep plants alive and healthy
- To make color
- To describe some important ideas in God's Word. In this unit's memory verse, light helps us understand the glory of the Lord.

What Is Light?

This is a question that not even Beamer can answer perfectly. Only God understands everything about light because He is so wise. Let's see what we do know about light:

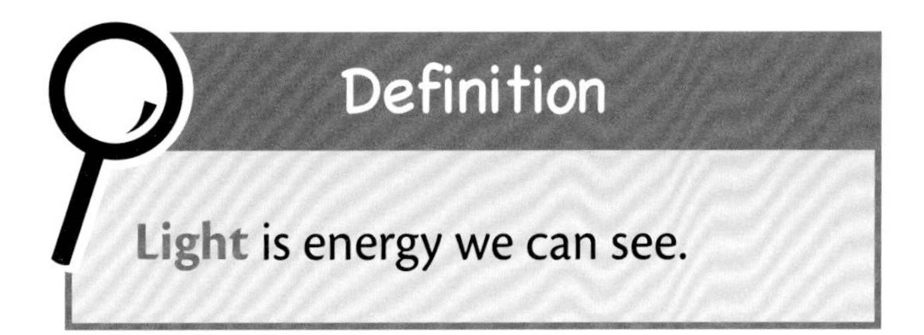

NASA's New Horizons Spacecraft

Light doesn't weigh anything. That means it isn't heavy when you try to pick it up. Light doesn't take up any space either, which means it doesn't push things out of its way when it shines into your room. Light is nothing but energy. Energy may be hard to understand. You can think of energy as being power. God has *all power* to do anything He wants. He also created many different kinds of power so that His creation would work and not fall apart. Some power He made is invisible, like the power that keeps us from floating off the earth. There are many kinds of power that we cannot see. But light is a power (or energy) that we *can* see.

When you see light, you are looking at energy. We know that light's energy travels very quickly. If you walk into a dark room and turn on the light switch, what happens? It seems like the light fills the entire room in a single second. Light travels from the light bulb to every corner of the room faster than you can even watch it move! Light moves faster than anything else we know. The speed of light is the fastest thing people have been able to measure. Your car drives very quickly on the highway. The fastest rocket ship

Light usually moves in a straight line.

is about 1,000 times faster than your car. But light is more than 10,000 times faster than that rocket ship. It took the fastest rocket ship 8 1/2 hours to leave the earth and travel past the moon.[1] Beamer could travel to the moon in about 1½ seconds![2]

We also know that light usually moves in a straight line. It cannot travel around corners unless it has help. The only way it can go around a corner is by bouncing off things that point it in another direction.

When you go into a dark room and turn on a light, the light instantly bounces off all the things in the room, zig-zagging everywhere. Some of it even goes into your eyes so you can see the things that it bounced off of!

Time to do Activity 1 in the Activity Book!

How Does Light Act?

People have studied light for thousands of years. But we still don't understand everything about it. We don't understand exactly how light works. People have learned that it acts like the waves you see when you throw a rock in a pond. Light moves in little waves just like water moves in little waves when your rock hits it.

But people have also learned that light moves in little pieces called **photons**. These photons don't act like waves. Light seems to act differently when we study it with different machines. We haven't figured out

Definition

Photons are the smallest pieces of light.

Light acts like the waves you see when you throw a rock in a pond.

everything about how light works. But God knows it all. He is the only One who understands everything about light.

Beamer has been pretty quiet about himself, but this would be a good time to talk about light beams. Light beams are very useful. Lighthouses use light beams to warn boats on the sea that they are near land. Sometimes we see beautiful sunbeams stretching to the earth from between clouds. They draw our eyes up to show us a beautiful sight that reminds us of heaven and God's glory.

When light shines through mist or dust in the air, it's easier for us to see the light beams. This is because the light bounces off the mist or dust that it's shining through. Some of the light waves in a lighthouse's beam bounce off the ocean's foggy spray and into our eyes. Thankfully, most of the light waves keep on their straight path and do their job brightly for the boats.

Time to do Activity 2 in the Activity Book!

Things look gray at nighttime without light's energy hitting them.

God Made Colors!

Yay! Now we get to learn the most beautiful thing about light. COLOR!

At nighttime, everything looks gray and black. Things really don't have any color by themselves until light's energy hits them. Then something beautiful happens!

Even though light looks white to us, it is made up of every color on earth. God made each thing in the world in a special way so that when light hits it, some colors bounce off and others are absorbed or soaked in. The colors that bounce off are the ones that we can see. If you get hungry in the middle of the

night, you might go into the kitchen and see a dim basket of fruit on the table. When you turn on the light, an apple grabs all the colors except red. Red bounces off, and some of this red color shines into your eyes. You think, "That apple is red." Maybe you hoped for a banana. Yellow bounces off the banana, so it looks yellow to you. If the banana is old and black, then it will absorb all the colors, so none will bounce back at you. That's why it looks so dark. If you use a paper napkin, it looks white because all of light's colors are bouncing off it.

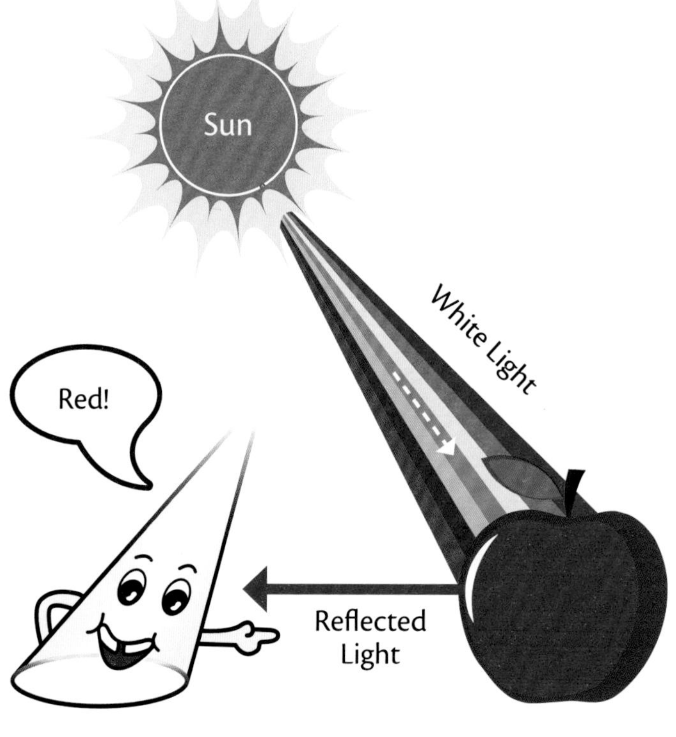

Crater Lake, Oregon, where the water is so pure and so deep (2000 ft, 600 m) that it absorbs almost all the colors except this unusual blue known as "Crater Lake blue."

[God] called you out of darkness into His marvelous light. (1 Peter 2:9)

Prayer

Thank You, God, for making our world beautiful and safe with light. Thank You for all the different colors. Amen.

Time to do Activity 3 in the Activity Book!

CHAPTER 2

Shadows, Reflections, and Bending Light

Do you remember that light usually travels in a straight line? If you go outside on a sunny day you will see bright places where the sun's light reaches the ground. But there will be a darker place near you where the sunlight cannot hit the ground. It is hitting you instead and can't go through you or around you. You are casting a **shadow**.

Shadows are not completely black. There is some light in shadows because light is bouncing off other things around us and hitting the shaded ground. Isn't God wise to make light bounce so that we can see, even in shadows? If we did see a totally black place on the ground, we would be careful. If there is nothing there for light to bounce off of, we are seeing a very deep hole!

Definition

A **shadow** is an area that is not lit because something is blocking the light.

Tree branches block light and cast a shadow.

Shadows under things

God's creation of light and shadows helps us understand the world around us. Artists draw shadows under things to keep them from looking like they are floating. In the picture (right), we can tell that one person is jumping and one is not because of their shadows.

Shadows on things

Shadows are not only *under* things. They are also *on* things. You might have a big shadow behind you, but you also have little shadows on your face and clothes. When light shines on one side of you, the other side of you will be shaded.

Shadows on things help us understand their shapes. Without shadows, things would look flat. Look at the circle, the ball, and the red blood cell. Their outlines are the same, but the shadows help us know their shapes.

God uses shadows to protect us. The shadows of trees protect us from the heat of the sun. The roof and walls of your home give you shade. God doesn't have a shadow, but He uses the idea of His shadow to teach us that He protects us:

He who dwells in the secret place of the Most High
Shall abide under the shadow of the Almighty. (Psalm 91:1)

Time to do Activity 4 in the Activity Book!

Reflection in smooth water

Reflection

Reflection is a word we use to talk about how light bounces off things. Light bounces or reflects very well off of things that are smooth and shiny. A mirror is the best place to see reflections, but water reflects things too. We can see many beautiful sights in God's creation because of the way water reflects. When water is smooth, the reflection is like a mirror. When water is rippled or rough, the water reflects things from many different directions.

Reflection in rippled water

**As in water face reflects face,
So a man's heart reveals the man.
(Proverbs 27:19)**

A car mirror shows the road behind.

Car mirrors help drivers stay safe by letting them see what is around them and behind them without having to turn their heads. They can easily know what is near them with only quick looks at their mirrors.

Time to do Activity 5 in the Activity Book!

When Light Bends

Light can bend when it travels through different temperatures. That means it may bend when it's moving from a hot place into a cold place or from somewhere cold to somewhere hot. This can happen at sunrise or sunset. Look at a picture of a sunrise or a sunset. (*Never* look at the real sun, even if it doesn't seem bright.) In pictures, you might notice that the sun looks squished when it is close to the earth. You are seeing the bottom of the sun that has already set, but its light is being bent back up where we can see it. The different temperatures near the earth's surface cause the sun's funny shape.

You can also see that different temperatures are bending light if you look far down a road on a hot day. If it looks like there is water on the road that disappears when you get closer, you have seen a **mirage**. The word *mirage* is like the word

Different air temperatures make the setting sun look squished.

mirror. When you see a mirage, you are seeing a reflection of the sky. Hot air above the road has bent some of the light from the sky and bounced it into our eyes. Since we usually see the sky's reflection in water, we are tricked into thinking there is water ahead.

Mirage on hot road

Light can also bend when it travels between air and water. God's colorful rainbows appear because of bending and bouncing light inside each raindrop. Rainbows can only be seen if the sun is behind us. The white light of the sun, with all the colors in it, comes from behind us and hits drops of water in the air. The drawing below shows how a water drop bends the light as it enters it and separates the colors a little. The colors bounce off the back of the water drop and come back toward us. As they come back into the air, the colors are separated even more to help us see them!

See how only the red hits the eye from the higher raindrop? Only violet hits the eye from the lower raindrop. The other colors coming from their raindrops are not shown, but the colors in between would be seen from raindrops in between the highest and lowest ones. This allows us to see a rainbow with red on the top and violet on the bottom.

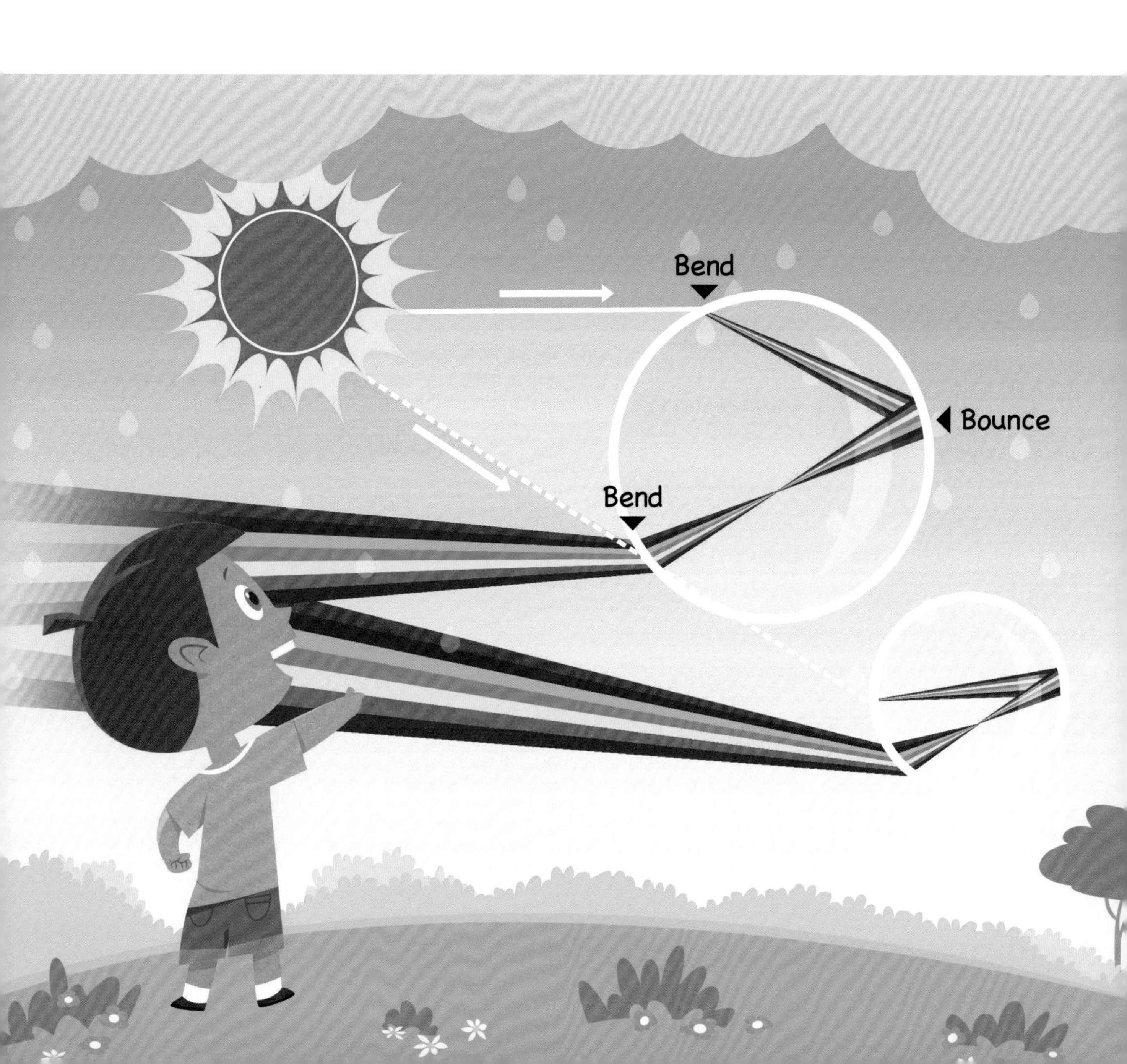

In the Bible, we read about a time when people became very wicked. God sent a flood over the whole earth to destroy all life except for Noah's family and some of every animal. He saved these in an ark. Afterward, God promised Noah that He would never flood the earth again. He gave us rainbows to help us remember His promise.

Raindrops bend light to make rainbows.

I set My rainbow in the cloud, and it shall be for the sign of the covenant between Me and the earth. . . . the waters shall never again become a flood to destroy all flesh. (Genesis 9:13, 15)

Prayer

Thank You, God, for Your beautiful world of light and shadow, shiny reflections, and the bending of light. We cannot touch a rainbow, or go where it seems to be, but we can praise You for making it so beautiful! Amen.

Time to do Activity 6 in the Activity Book!

CHAPTER 3

The Sun and Other Kinds of Light

The Sun

God made our sun to be Earth's greatest giver of light. Every minute of every day, the sun is shining on the earth. But we don't always see the sun shine. That's because Earth is a spinning ball. Every 24 hours, the earth spins around once. When we wake up each morning, our part of Earth has just spun into the sun's light. When we look outside during the day, it might seem like the sun is moving across the sky. But the sun is really staying in the same place. We are the ones who are moving. All day long, Earth is spinning.

The sun seems to move across the sky as we spin. Night comes when our part of the earth spins away from the sun and we are in Earth's shadow. Our part of the earth remains dark until we spin back into the sun's light the next morning.

In [the heavens] He has set a tabernacle for the sun,
Which is like a bridegroom coming out of his chamber,
And rejoices like a strong man to run its race.
Its rising is from one end of heaven,
And its circuit to the other end;
And there is nothing hidden from its heat. (Psalm 19:4-6)

Time to do Activity 7 in the Activity Book!

All day long, we see light shining down on us from the sun, but there are other kinds of energy coming from the sun that we cannot see. One of those is heat.

God did not make the sun solid like the earth. He made it out of very hot gases. The sun's temperature is close to 11,000 °F (6,000 °C)! Its heat makes Earth warm enough for us to live. God put our earth at just the right distance from the sun. If Earth were closer to the sun, it would be too hot for anything to stay alive! Everything on Earth would burn up or melt. If Earth were farther away from the sun, things would freeze, and everything would

Vitamin D is a special vitamin that our skin makes when we stand in the sun. People need 10–30 minutes of sunshine on their skin every few days to make enough Vitamin D. You can get all you need in half the time it takes to get a sunburn. So, go out and enjoy God's light for a little while!

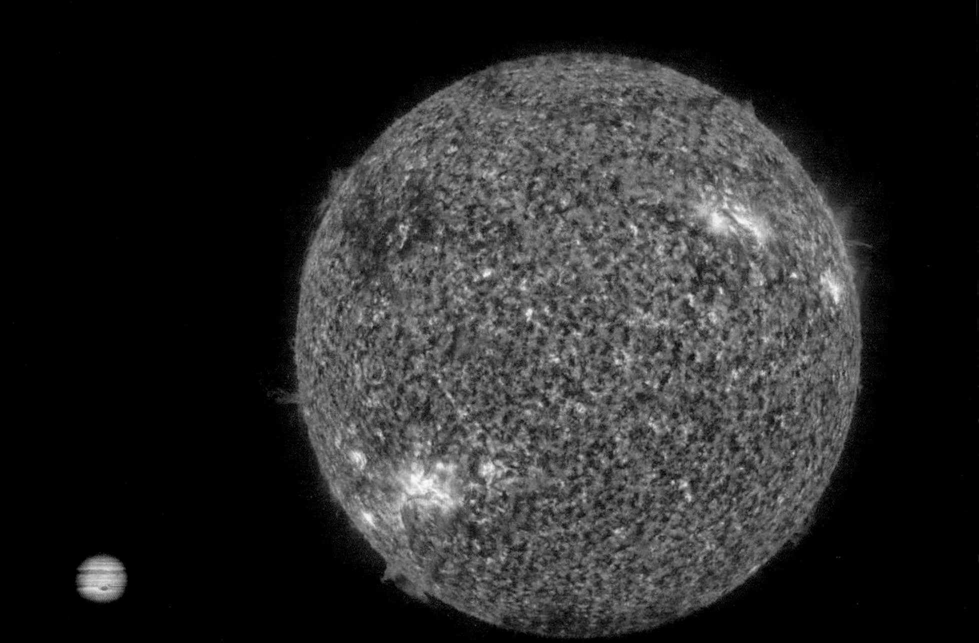

God gave the sun the perfect amount heat for us to live on Earth.

die. God put Earth in just the right place so that we could live on it. He is a very wise God. He knows exactly where Earth needs to be so that we can live. He cares for us!

The sun is very big. More than a million earths could fit inside it! The sun is also very far away. Light takes about eight minutes to travel the 93 million miles (150 million km) from the sun to the earth. It would take a car 150 years to drive that far if it drove at fast highway speeds and never stopped!

Other Natural Light and Heat

Heat from the sun warms the whole earth, but there are smaller sources of heat and light that God made right on the earth.

Fire has always been a useful way to get light. Fire is especially helpful because of the heat it gives off. There are many parts of the world where people could not live in the winter if they didn't have fire. But in order to have fire, you must have something to burn. We call this **fuel**.

Beamer's List of Fuel

1. Wood
2. Charcoal
3. Peat moss
4. Wax
5. Coal
6. Natural gas
7. Cow and buffalo manure
8. Oil from under the ground

Definition

Fuel is something burned to make heat.

God gave us fuel to warm our homes.

Fireflies give off light without heat.

God also made some living things that give off light without heat.

Fireflies are beetles that can quickly turn light on and off inside their bodies. They do this to find each other so they can make eggs that will become more fireflies. How do they make this light? Inside their bodies is something that glows when it is mixed with other things. This makes the light turn on. The fireflies can turn their lights off by mixing in something else. Different kinds of fireflies have different ways of blinking so that the same kinds can find each other.

Anglerfish

Anglerfish live deep in the ocean where little or no light can reach them. God has given them fishing poles on their heads. Right in front of their sharp teeth is

Tiny glowing animals light up waves.

where the hook would be. But instead of a hook, there is a glowing blob. This light attracts other fish. The fish think the light must be something good to eat. Oops! When they try to take a bite, they are eaten instead.

Foxfire is a type of glowing mushroom found in rotting wood.

Ocean waves can glow when there are millions of tiny glowing animals gathered there.

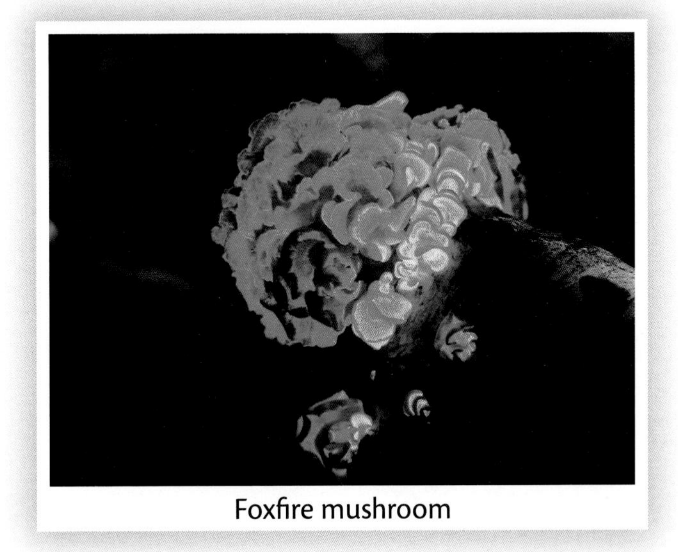

Foxfire mushroom

Time to do Activity 8 in the Activity Book!

People Use Light

God is the best Inventor. An inventor is the first one to think of and make something. In the beginning, God thought of everything and then created it. When people invent things, they do this by studying God's creation. They learn about something God made, and then they make something useful that no person has made before. Here are some things that inventors have made by studying light and heat:

- People invented ways to make things look bigger by bending and reflecting light. Some people invented telescopes to see large objects that are far, far away. Others invented microscopes to see small things up close.

Looking at the moon through a telescope

Looking at tiny creatures through a microscope

- People invented ways to use fire for light.

Candles

Ancient oil lamp

Kerosene lantern

- People invented ways to use electricity to make light. Electricity makes something glow inside light bulbs and tubes.

Light bulbs

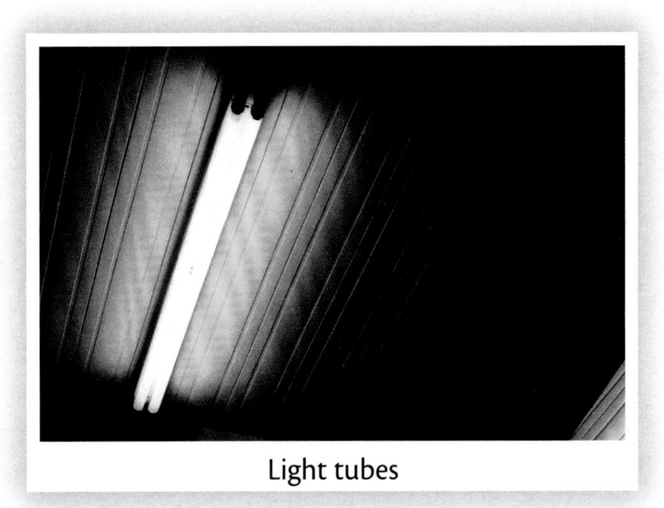
Light tubes

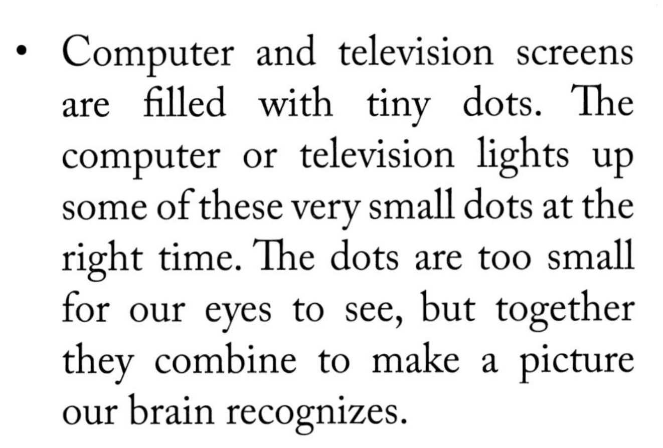

- Computer and television screens are filled with tiny dots. The computer or television lights up some of these very small dots at the right time. The dots are too small for our eyes to see, but together they combine to make a picture our brain recognizes.

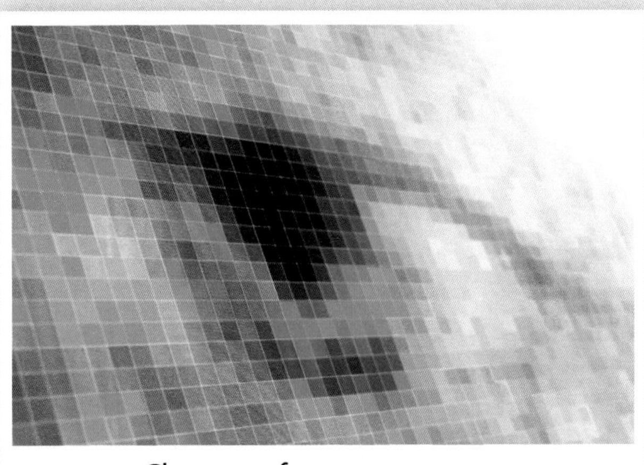
Close-up of computer screen

Panels collect sunlight for solar energy.

- People studied reflection and made fiberoptic cables to carry light and lots of information farther than before. But how do these cables carry light? Even though light travels in a straight line, reflection lets it travel through bent cables. The inside of a fiberoptic cable is made of glass walls. This glass is as thin as a person's hair, but it still reflects light. The light bounces off these glass walls and travels speedily through the cable. These cables are covered with a special plastic so that the light doesn't break through. Instead, it travels down the entire cable, bouncing back and forth off the mirrors as it goes.

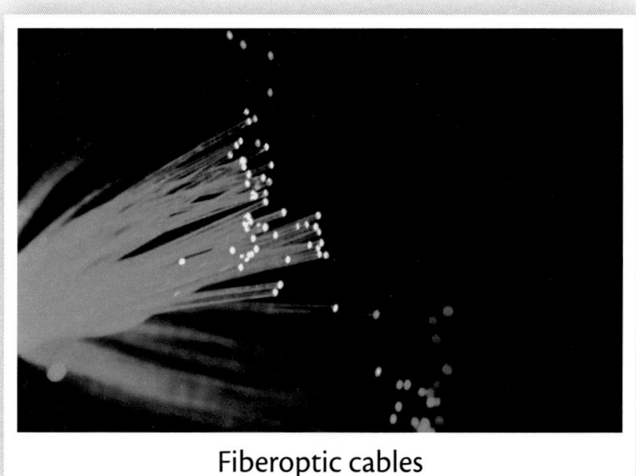

Fiberoptic cables

**For You will light my lamp;
The LORD my God will enlighten my darkness.
(Psalm 18:28)**

- People invented a way to store solar energy so they can use the warmth from the sun for electricity.

Prayer

Thank You, God, for the sun that gives warmth and light. Thank You for giving people Your wisdom to invent ways to use light. We praise You for your care! Amen.

Time to do Activity 9 in the Activity Book!

CHAPTER 4

The Eyes Welcome Light

Then God saw everything that He had made, and indeed it was very good. (Genesis 1:31)

God said that everything He made was very good. He gave us eyes so we would be able to see the good things He made. Eyes are so complicated that we still don't understand everything about them. Here are some amazing things about our sight:

Your eyes can see things close to you and far away.

- Human eyes are so powerful that they could see a candle in the dark from a mile and a half away.[3]
- Our eyes focus on fifty different objects every second.[4]
- Your eyes can tell the difference between 10 million colors.[5]

- Most of our memories come from what we have seen with our eyes. That means our eyes help us learn![6]
- It took about 40 weeks for you to grow inside your mother before you were born. Your eyes started to form only two weeks after you began growing![7]

Time to do Activity 10 in the Activity Book!

How Sight Works

The black circle in the middle of the eye is the **pupil**. It looks black because we are seeing the inside of the eyeball where light is not reflected out again. The pupil is covered with clear skin. Light goes into the eye through the pupil.

The colored circle around the pupil is the **iris**. When we talk about what color our eyes are, we are talking about the color of the iris. The stripes in the iris are little muscles that open and close the pupil to let the right amount of light in. In darkness, the pupil opens wider to let more light in. In bright light, it closes a little to protect the eye from too much light.

Three things work together for us to see: our eyes, our brains, and light. If one of these is not working, we cannot see.

God made your amazing eyes!

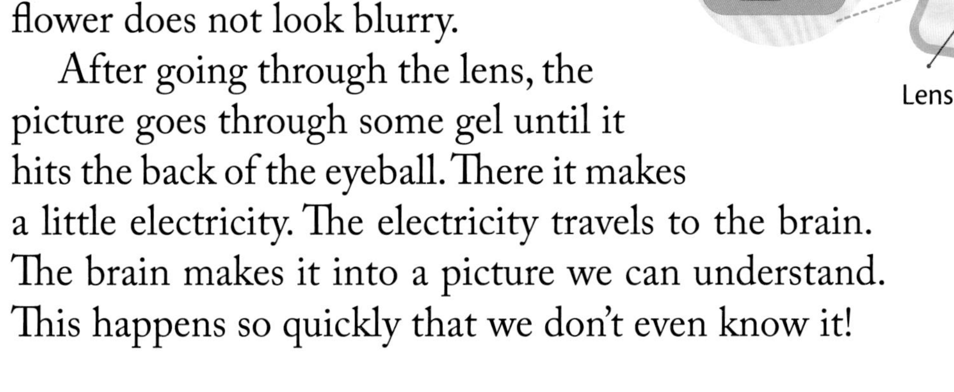

In this picture, light is bouncing off the mountains and traveling through the pupil. Then the light goes through the **lens**. The lens changes shape to keep the far-away mountains from looking blurry. If we are looking at a flower close to us, the lens changes to a different shape so that the flower does not look blurry.

After going through the lens, the picture goes through some gel until it hits the back of the eyeball. There it makes a little electricity. The electricity travels to the brain. The brain makes it into a picture we can understand. This happens so quickly that we don't even know it!

Eyes + Light + Brain = Sight

Isn't God amazing to make our eyes able to use light and our brains to see His beautiful work? Only God could do this! It could not happen by accident.

Why God Gave Us Two Eyes

You might wonder why God gave you two eyes instead of just one. Why do we need two?

Two eyes help us see better. With only one eye, we would have trouble seeing how far away something is. Because our eyes are on different sides of our face, they each see a slightly different picture. Our brain helps us use the two pictures to find out how far away something is. This is very important to know if you're crossing a street or trying to catch a ball. Having two eyes keeps us safe so we don't bump into things.

Time to do Activity 11 in the Activity Book!

Eyes of Creatures

God made the eyes of most of His creatures differently from ours. Each kind of animal has eyes just right for how God made it to live.

Animals that need to hunt in the dark, or those that need to protect themselves from other animals in the dark, have a special covering inside their eyeballs that reflects light. As light goes into their pupils, it bounces around the inside of their eyes and lets the animals see things better in the dark. We can sometimes see the light bouncing back out of their eyes.

Definition

The **tapetum** is the layer inside the eyes of some animals that causes light to reflect inside their eyeballs. This helps them see better in the dark.

A snowy owl hunts well.

Animals that hunt have eyes in the front of their head. This helps them hunt. Because both of their eyes are looking at the same thing, they can tell exactly how far away it is.

God gave animals with hooves a different kind of eye. These animals don't hunt, but they must watch for enemies in all directions. Their eyes are on the sides of their heads and usually stick out. This helps them watch for danger. God gave them pupils shaped like rectangles so they can see farther forward and backward at the same time.

A horse's rectangular pupils

Some animals have round pupils. This helps them see far away to hunt for food.

Bald eagle

Wolf

Animals that need to sneak up and attack their food have pupils that look like tall slits.

Crocodile

Cat

Monkeys see the same colors you do.

Monkeys and apes are the only animals that see color as we do. Most animals do not see red or colors made with red, like pink and orange. The blue, green, and yellow they do see is usually faded. Most of these animals see much better in the dark than we do.

Birds see more colors than you do.

It would be fun to see like a bird for a day! Birds see all the colors we do plus an extra one: ultraviolet. We can't see this color, but many flowers and birds have ultraviolet designs. Birds use these designs. Some birds may be able to find each other by the ultraviolet marks that we (and other animals) cannot see. Birds that hunt may find their food more easily by seeing the ultraviolet color of fur and urine left by another animal.

Let's compare you to an earthworm!

You

- Have eyes to see beauty, to help you learn, and to keep you safe.
- Live in the light except when you're sleeping.
- Can see your colorful food.
- Can thank God for worms because they turn rotten plants into good soil.

A Worm

- Has no eyes. Instead, its skin can tell when it's in the light. The skin warns the worm so it can crawl back underground to keep from drying up.
- Lives in darkness all the time.
- Crawls blindly through its food (rotten plants), eating a tunnel to travel through.
- Cannot thank God like you do.

Praise God! He gives us eyes that are perfect for what we need. And He gives worms exactly what they need too, even though they don't have eyes like we do.

Prayer

Thank You, God, for our amazing eyes that see babies and books up close, and trees, hills, and the moon far away. Thank You that my eyes are in the front of my head and protected by You. Please help me to take care of my eyes and the sight You gave me. Amen.

Time to do Activity 12 in the Activity Book!

UNIT 2

The Earth

I like to shine on the whole earth, every day, summer and winter! I like to shine on land and sea, and rocks and dirt. This memory verse is about Jesus being the light of the world. He doesn't only shine on the earth like I do, but He is also a light for people's souls. He can take away the darkness of sin.

Jewels are precious stones that come from inside the earth. Our hymn says that children who love Jesus are His special jewels.

Memory Verse

"I have come as a light into the world, that whoever believes in Me should not abide in darkness."

(John 12:46)

Hymn Singing

Jewels (When He Cometh)

When He cometh, when He cometh
To make up His jewels,
All His jewels, precious jewels,
His loved and His own.

Chorus:
Like the stars of the morning,
His bright crown adorning,
They shall shine in their beauty,
Bright gems for His crown.

He will gather, He will gather
The gems for His kingdom,
All the pure ones, all the bright ones,
His loved and His own.

Little children, little children
Who love their Redeemer,
Are the jewels, precious jewels,
His loved and His own.

You can listen to this hymn by searching for "Jewels (When He Cometh)" on the internet.

As they travel around the earth, man-made satellites gather information about the earth and space or help direct phone and television signals.

CHAPTER 5

The Earth Is Big

He has made the earth by His power;
He has established the world by His wisdom.
(Jeremiah 51:15)

Too Big to Walk Around!

Do you know how big Earth is? If you went into your front yard and looked around, you might not be able to see very far. It might not seem like Earth is very big. That's because trees and buildings get in the way. But if we go to an open field or a beach or to the top of a high hill, we can see much farther. Then, if we look far into the distance, it looks like the sky and earth touch each other. This is called the **horizon**.

If you are six years old and are standing at the edge of the ocean, the horizon would be about 2 ½ miles (4 km) away. That means you could see 2 ½ miles of ocean in front of you! If you were standing on a boat in the middle of the ocean, you could see 2 ½ miles of water in every direction.

> **Definition**
>
> The **horizon** is the line where we see the sky and the earth meet.
>
> The **equator** is an imaginary line that stretches around the middle of the earth.

Earth is shaped like a big ball. It is smaller at the top and bottom and widest in the middle. The top and bottom of the earth are called the North Pole and the South Pole. When people make a globe or a map of the earth, sometimes they draw a line around the middle of it. This line is called the **equator**. Our Earth might not look very big in a picture or on a map. But if you wanted to walk all the way around the widest part of it, you would have to walk five miles a day for 13 years! That's a long time!

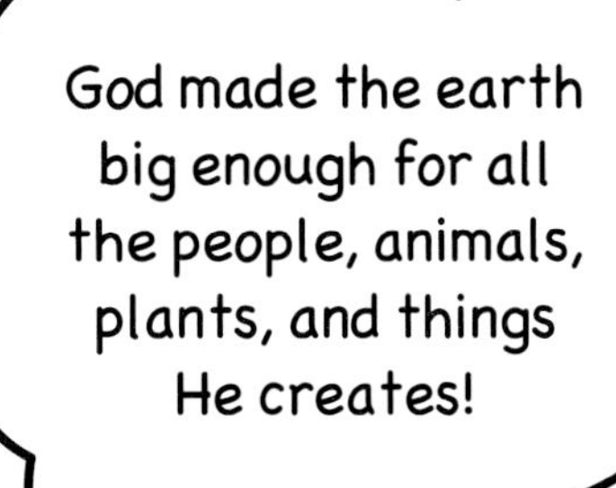

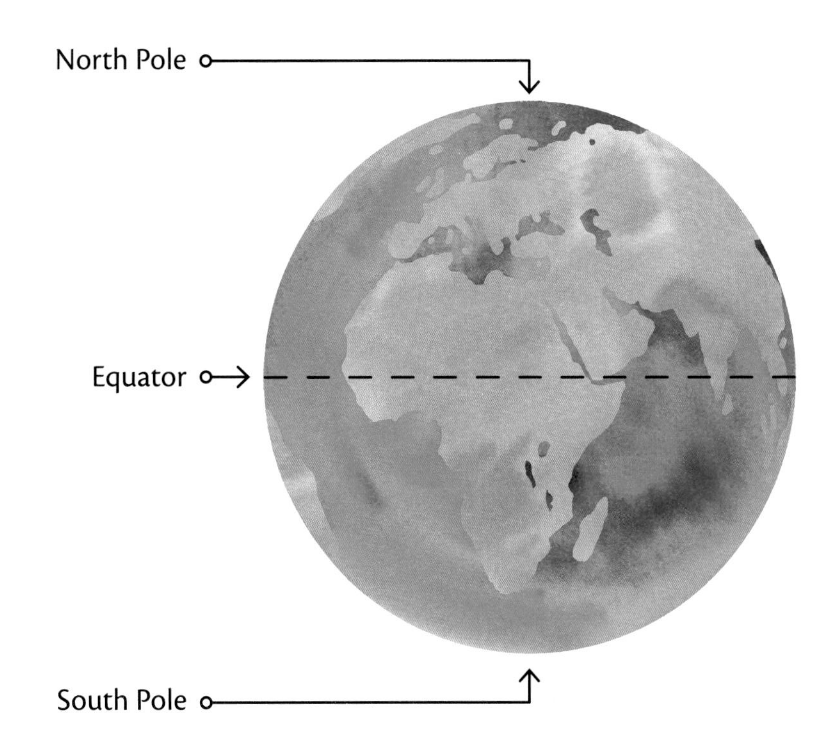

Time to do Activity 13 in the Activity Book!

Too Heavy to Weigh!

If you tried to pick up the earth, it would be very heavy! Inside it are layers of very heavy things.

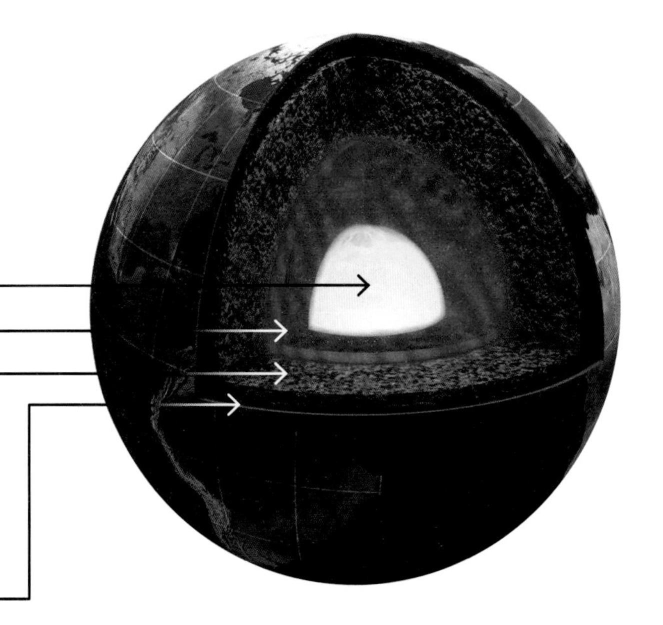

- The inner core is made of very hot, hard metal.
- The outer core is made of very hot, melted metal.
- The mantle is made of very hot, melted rock.
- The crust is made of cool, hard rock. We live on the crust.

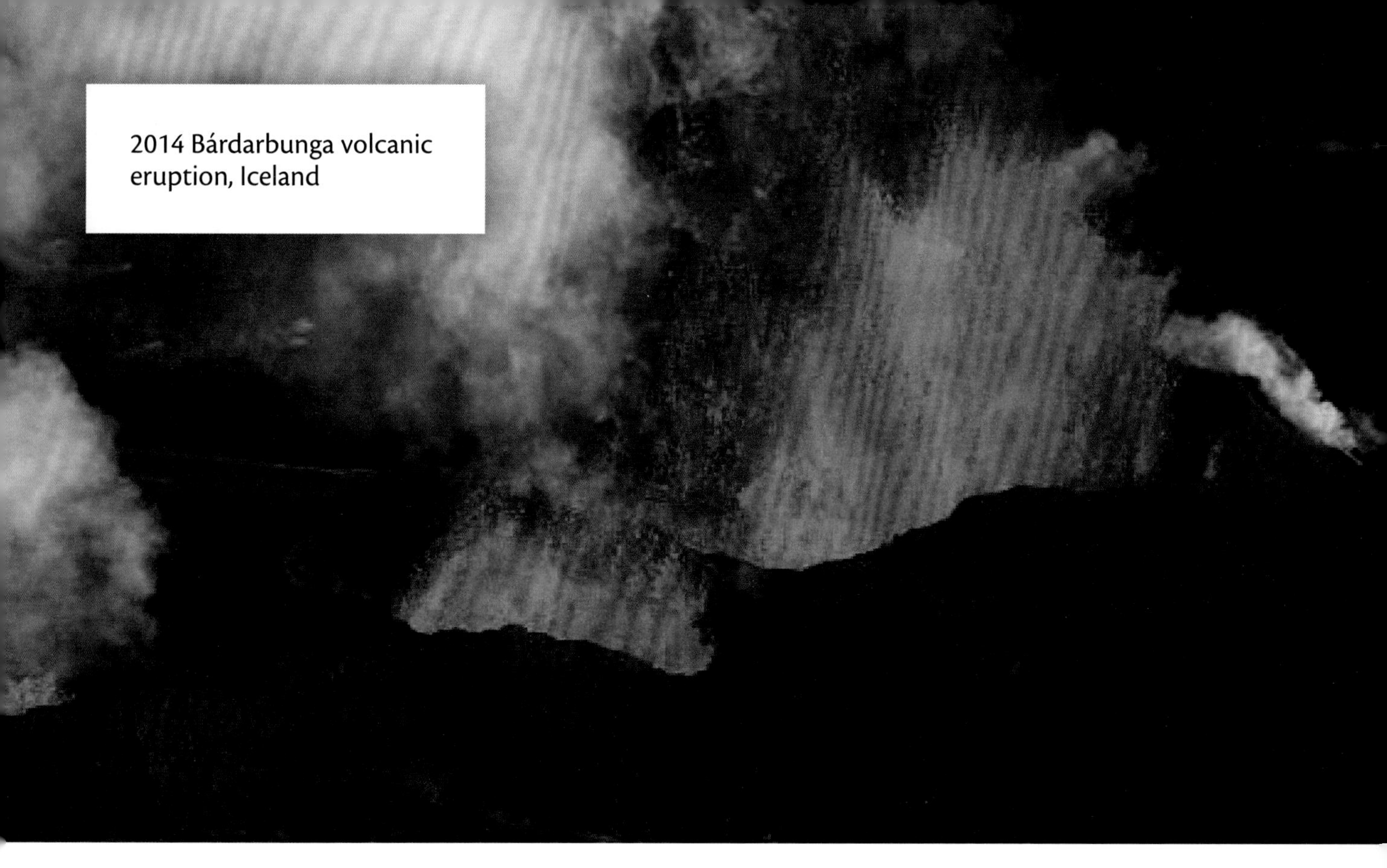
2014 Bárdarbunga volcanic eruption, Iceland

Have you ever seen a picture or watched a video of a volcano erupting? With some volcanoes, hot lava explodes out the top and flows down the sides. Do you know what lava is and where it comes from? It is melted rock from Earth's mantle. When melted rock breaks out of a volcano, we call it **lava**. But if the melted rock stays inside the earth, we call it **magma**.

Some volcanoes have lava that flows instead of exploding.

Time to do Activity 14 in the Activity Book!

It Pulls So Hard!

God has made an invisible pull that keeps us and all things on the earth. We call this pull **gravity**. Without it, we would all go floating off into space. The room above seems to be missing that pull!

> **Definition**
>
> **Gravity** is the energy that pulls things together.

Where does gravity come from? We don't understand everything about it, but we know that gravity is here because of the earth. The earth's gravity is like a magnet. It pulls things to it. If you jump into the air, you'll fall back down to Earth. Why? It's because Earth's gravity is pulling you down. If you throw a ball into the air, Earth's gravity will pull it down too.

Big things in space, like the sun, have a lot of gravity. Little things in space, like the moon, have a small amount of gravity. The amount of gravity depends on how big something is. God made Earth the perfect size to give us the right amount of gravity. What do you think would happen if God had made Earth a different size? Let's look and see.

If Earth were smaller

- We would float off into space.
- We would not have air to breathe because there would not be enough gravity to keep it here.
- We would not have water to drink. Water needs air pushing on it to stay liquid.
- We would get sunburned very quickly. Air keeps out some sunlight, and gravity keeps our air here.

If Earth were bigger

- We could be crushed by too much gravity.
- We would get too hot. Too much gravity would keep too much air here. Extra air would keep too much of the sun's heat close to the earth.
- We might have poisonous air to breathe. Too much gravity could keep the wrong kind of air close to the earth.
- It would be very hard to lift our legs and walk with extra gravity pulling them down.

Astronaut floating in space where there is not enough gravity to pull him to Earth

People do not understand everything about gravity. We know what it does and how much it pulls, but no one knows how it does that. But God knows! We can be glad that Jesus keeps all things where they should be.

[Jesus is] upholding all things by the word of His power. (Hebrews 1:3)

Earth's gravity is pulling everything toward the center of the earth. That's why, no matter where we are in the world, we feel right-side up.

Prayer

Thank You, God, for making our big earth just the right size for life! Thank You for the interesting lava that comes out of the mantle through volcanoes. We praise You for holding everything where it should be. We can trust that gravity never changes. Help us to remember that You never change, and that we can trust You. Amen.

Time to do Activity 15 in the Activity Book!

CHAPTER 6
The Crust

And God called the dry land Earth, and the gathering together of the waters He called Seas. And God saw that it was good.
(Genesis 1:10)

The Rocky Crust

Wherever we are on the earth, either on dry land or deep in the seas, the earth's crust is underneath us. It is good! God made it just right.

The crust under dry land is 20-28 miles thick (30-45 km). It is made mostly of rock with a thin layer of dirt on top. Sometimes, if the dirt has been washed or blown away, we can see the solid rock underneath it.

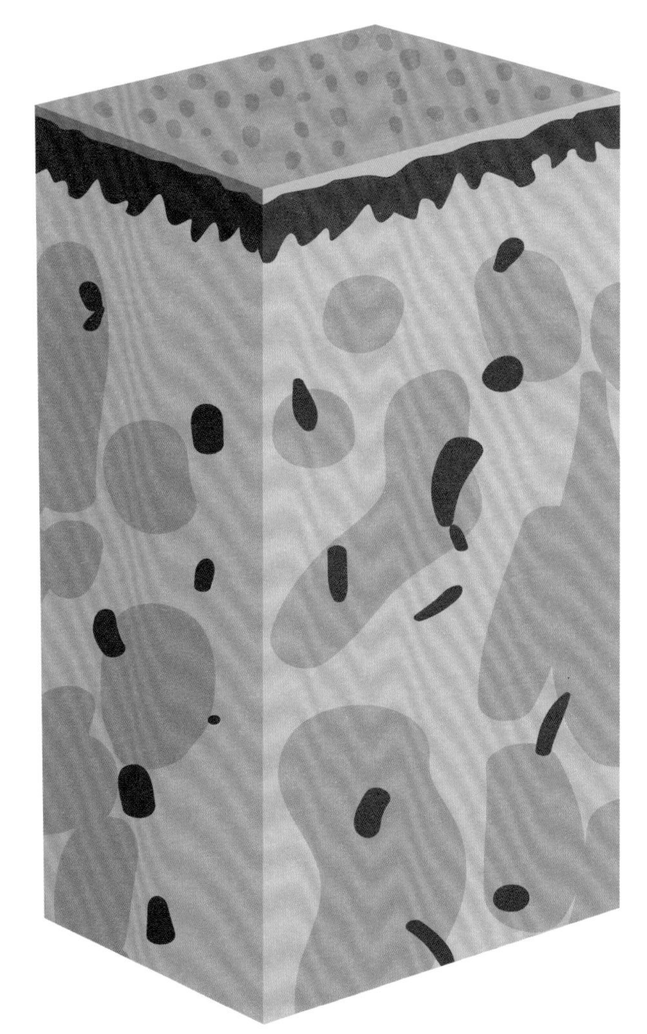

God has put a thick layer of rock under Earth's plants and dirt.

There are three kinds of rock in the crust:
Igneous—hot magma that has cooled and hardened. Obsidian and granite are two types of igneous rock.

Obsidian

Granite

Sedimentary—tiny pieces of rock, such as sand, that piled up underwater and got stuck together. Sandstone and limestone are two types of sedimentary rock.

Sandstone

Limestone

Metamorphic—rocks that were changed into different rocks by being heated very, very hot or by being smashed and squished together. Marble and slate are two types of metamorphic rock.

Marble

Slate

Almost all of Earth's crust is made of igneous rock, but we don't often notice it because other kinds of rock are usually on top of it.

God made Earth's crust out of rocks. There are many useful and beautiful things that come from rocks. The most useful thing is dirt. Most dirt is made of ground-up rock. Most plants need dirt to grow. If they don't have it, they will die. Jesus told a story about seeds that landed on bare rock and could not grow into plants. Plants need dirt to protect their roots, to store water, and to give them minerals to help them grow.

Time to do Activity 16 in the Activity Book!

God Put Treasures in the Earth

Metal is another useful thing that comes from rocks. It can be beautiful too. We can use metal for many things.

- Metal is shiny. The shiniest metals are used for jewelry and coins.

Silver and gold jewelry

Silver and gold coins

- Heat and electricity travel easily through metal.
- Metal can be pounded and stretched. A gold nugget the size of a baby's fingertip can be pounded into a square sheet of gold that would be as tall as your waist.

Copper Wire

Metal can be pounded and stretched.

- Strong metals are used for making cars, trains, ships, and buildings.

Cars, trains, and buildings made of metal

Steel cables on a bridge

In Earth's crust and cores, we find a lot of a metal called **iron**. One third of the earth is iron! People have used iron to invent another very strong metal called **steel**. People make steel by mixing different metals with iron.

Jewels are very beautiful rocks. They are found buried deep in the earth. You will probably never find a jewel. If you bought one, it would cost a lot of money. It would be a treasure! Jewels are very hard. Diamonds are the hardest things in the world. Jewels are very clear. You can see right through them! Jewels have edges, points, and flat sides.

Jewels were made deep in the earth where it is very hot and where there is a lot of pressure that pushes on the rock or magma around them. Jewels used cracks in hard rocks or bubbles in magma as roomy places to grow. Different jewels were made in different ways.

Only God understands how diamonds were made, but we know they were made in the earth's mantle and were then brought up to the surface by exploding volcanos.[8]

Uncut diamond

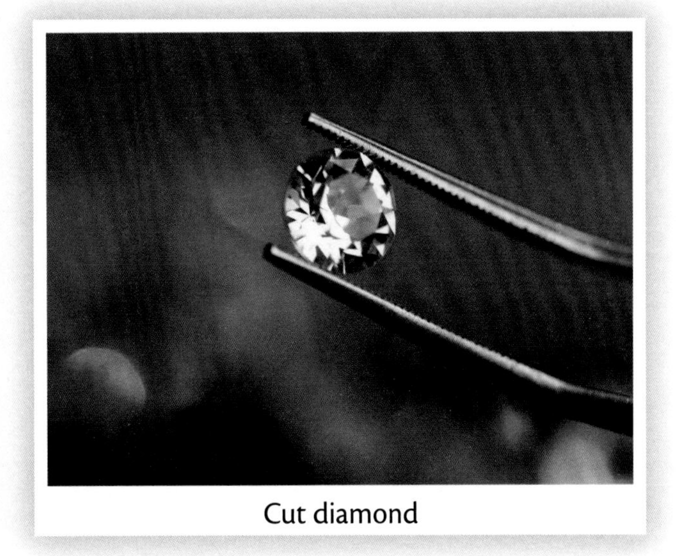
Cut diamond

Rubies are red jewels that are mostly found where two big pieces of land (Asia and India) crashed into each other and made high mountains at the time of Noah's flood. Rubies are not found in very many places because they could not grow if iron was near them.

People cut and polish jewels to make them as beautiful as possible.

Uncut ruby

Cut ruby

Time to do Activity 17 in the Activity Book!

The Changed Crust

We don't know what the earth's crust looked like right after God created it. We do know that the huge flood God sent when Noah lived made a lot of changes.

God protects Noah's family while the flood changes the earth.

[During the flood,] the world that then existed perished, being flooded with water. (2 Peter 3:6)

Today, we see mountains that were pushed up when big pieces of land crashed into each other during the flood.

We also see thick layers of rock and sand that settled at different times. We call these **sedimentary layers**. God probably made some sedimentary rock when He made the dry land at creation. But more sand was made during the flood as rocks pushed and ground against each other. As this sand settled on the ocean floor, it made layers of sand. The ocean water moved in different directions and sorted each kind of sand. As time passed, this sand hardened into rock. But sometimes these rock layers got bent before they could harden completely. This happened when the earth's crust moved or shifted.

In some places, we find canyons in Earth's crust. A **canyon** is a large, deep groove in the ground. Canyons were made where sand had settled but was washed away before it could harden. All over the world, we find canyons. Some of them are big, and some are small. Besides the Grand Canyon in Arizona in the United States, there is a very large canyon hidden under the ice near the South Pole. People have found other big ones at the bottom of the ocean.[9]

Mountains that were pushed up

Bent rock layers

Grand Canyon

Layered mountains

In other parts of the world, we find sedimentary rock layers that got hard but were broken as the crust lifted and left them at an angle.

We see sandstone that has rocks mixed in with it. This probably happened when the flood waters churned like water does in a washing machine.

We see mountains that were once volcanoes. The cracking of Earth's crust

Rocks in sandstone

Volcanic mountain

at the time of the flood let magma through and formed volcanoes.

We see **fossils** of dead animals and plants. They were killed in the flood and were buried in sand before it turned to rock.

Definition

A **fossil** is something that died and was buried and turned into stone before it could rot.

Dinosaur fossil

Petrified tree

Prayer

We see Your power, God, when we look at the earth that changed because of the flood. Thank You for saving Noah and his family. Thank You for making beautiful jewels and metals in the broken crust. Amen.

Time to do Activity 18 in the Activity Book!

CHAPTER 7

Water

The Ocean

You rule the raging of the sea;
When its waves rise, You still them.
(Psalm 89:9)

The ocean covers more of the earth than dry land does! This picture shows different views of the earth as it spins for one day. You can see how much of the earth is covered with water. Yet the Bible says that God has measured all the water in His hand!

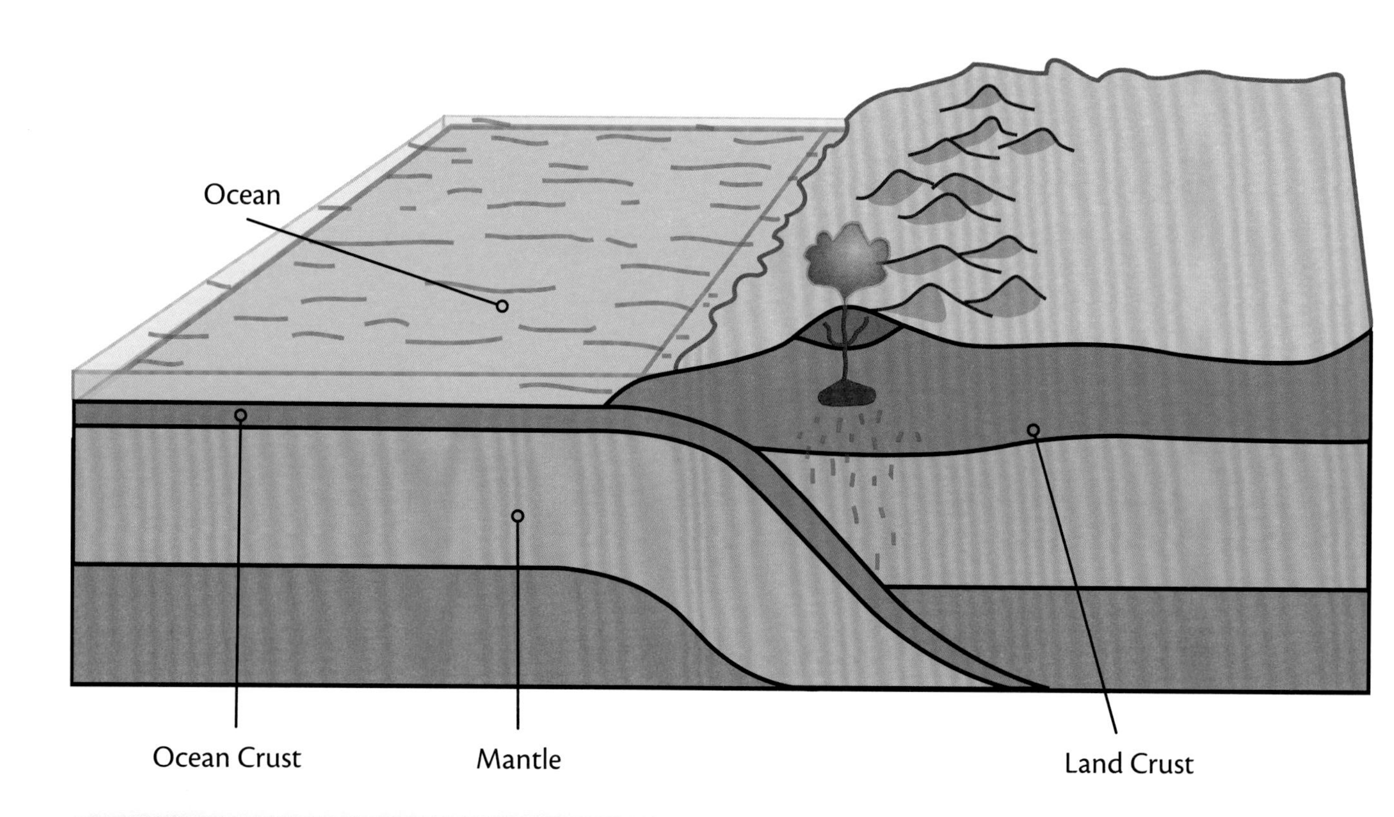

Dangerous wave caused by an earthquake

The earth's crust is thinner under the ocean than it is under land. Under the ocean, it is about four miles (6 km) thick. This crust is made of a heavier kind of rock than the rock under land.

Near land, there are places in the ocean where the crust is cracked. The heavy ocean crust slides under the lighter land crust. Around the time of the worldwide flood, it slid quickly. Today it is sliding only about as fast as your fingernails grow. Sometimes, pieces of sliding crust get stuck and then suddenly let go, causing an earthquake. If an earthquake happens in the ocean, it can cause a huge, dangerous wave.

The waves we see coming onto the beach one after another are made by wind. Wind can cause waves to crash onto the shore even if the wind is far, far away from shore. If the wind is blowing somewhere, it makes the water start to move up and down. That motion can travel for thousands of miles.[10] As the waves get closer to the beach, the deepest water is slowed down by the sand underneath it. The water on top moves faster, making a curling wave. It's the same thing that happens when you run and trip on something. Even though your legs try to move quickly to catch up with your body, the top of you is still going too fast, and you fall over!

Curling wave

God has made many things that work together so the ocean does its job well:

- Ocean water moves around the world. It doesn't sit still in one place. Instead, it swirls around and travels from one part of the world to another. The Bible says there are "paths of the sea" (Ps. 8:8) where the water moves. Moving water keeps the ocean mixed so that its plants and animals get the food they need. Moving water also keeps the ocean from getting too warm, too cold, or too dirty in one place. Ships use the paths of the sea to help them travel faster.
- Sea animals clean the water. One oyster can clean fifty gallons (190 L) of water a day!
- The ocean doesn't get very cold. Sunlight heats water near the

Rocks covered in oysters

Ocean plants

surface. Deep down, where Earth's crust is cracked and magma comes out, the water becomes as hot as fire. The paths of the sea spread this warmth around the world. This warmer water keeps the sea creatures alive.

Heated water coming up from ocean floor

- Ocean plants make air that we can breathe. They make more than half of the kind of air in the world that animals and people need to stay alive.
- The ocean always has the right amount of salt it needs for the things that live in it. Sometimes more salt comes into the ocean because of rain. When it rains, little bits of salt from the dirt wash into the water and make their way to the ocean. But the water doesn't get too salty because God put special animals and rocks in it that get rid of the extra salt.[11]

Time to do Activity 19 in the Activity Book!

Water Has Different Costumes for Different Jobs

Water is very important for life on Earth. If there was no water, there would be no people, no animals, and no plants. God made water very special so it could do its job well. He made water in such a way that it can turn into three different things. These things are **liquid**, **solid**, and **gas**.

Water washes things. It also cleans us inside. It travels around in our bodies and collects what we don't need. Then it takes that waste out when we use the toilet.

Definition

A **liquid** is something that can be poured and can run downward. When we say *water* we usually mean liquid.

A **solid** is something that is hard and has a shape. When water is solid, we call it *ice*.

A **gas** is something that is separated into such tiny pieces that it floats in the air. When water is gas, we call it *vapor* or *steam*.

Water's three costumes

Water can sometimes travel against gravity. This happens when it travels up from a plant's roots to its leaves.

When water freezes, it floats. This is very important for fish and other water creatures. Lakes that freeze are only frozen on top. This ice keeps the water underneath it liquid so that fish can stay alive during the winter.

Water vapor collects in the air to make clouds. Clouds are blown around in the sky and then begin to rain on the earth.

A girl waters the dirt so her plant's roots can drink.

Boy ice fishing

Time to do Activity 20 in the Activity Book!

The Water Cycle

Sometimes things happen in a certain order and end where they started. If this happens over and over, it is called a cycle. A baby eats, then sleeps, then wakes and cries because he is hungry, so he eats again. This goes on in the same order every day. We could call this a baby cycle. The earth has a cycle too. It has a cycle for day and night.

Water also has a cycle:

1. Rain falls on the land.
2. The extra water runs into lakes or the ocean, but the ocean does not get deeper!
3. Some of the water becomes vapor.
4. The vapor collects and makes clouds.
5. When the water in the clouds gets too heavy to stay up there, it falls as rain.

Water is cleaned as it travels through dirt. That sounds funny, but the dirt catches germs. Water is also cleaned when it comes out of the ocean as vapor because it leaves behind salt and other bad stuff for the ocean to take care of.

"Behold, God is great, . . .
For He draws up drops of water,
Which distill as rain from the mist,
Which the clouds drop down
And pour abundantly on man."
(Job 36:26–28)

Prayer

Thank you, God, for water in all its costumes. When I'm thirsty, help me think of Your eternal life. I'm glad that water goes all through me, cleaning me out and helping me live. I am glad water cleans everything. You are great and wise in making ways for water to be cleaned in the ocean and in the water cycle. Amen.

Time to do Activity 21 in the Activity Book!

CHAPTER 8

The Year and Its Seasons

What Is a Year?

What day is it today? Now think back a whole **year** to this day last year. You are one year older than you were last year at this time. Next year, you will be one year older than you are now. The earth will be one year older too. In that one year, the earth will travel all the way around the sun and will come back to the same place again.

> **Definition**
>
> A **year** is the amount of time it takes the earth to travel once around the sun.

Where do days and years come from? Remember, the earth spins around once each day. When our part of Earth is facing the sun, it is daytime. When we spin out of the sun's light, it is night. As the earth spins, it also travels around the sun. It spins around 365 times as it takes one trip around the sun. That means there are 365 days in one year. When you have a birthday, you have to wait 365 days until your next one!

How do we know how many days are in a year? People learned this by studying the sun's place in the sky. As God created the sun and moon and stars on the fourth day, He said:

"Let there be lights in the firmament of the heavens to divide the day from the night; and let them be for signs and seasons, and for days and years."
(Genesis 1:14)

One reason why God made the sun, moon, and stars is so people can keep track of days and years!

If you could look down on Earth from above the North Pole, you would see the earth spinning in a counterclockwise direction. This means that Earth is not spinning the same direction as the hands do on a clock. Instead, it is spinning the other way.

The earth also travels around the sun in a counterclockwise direction.

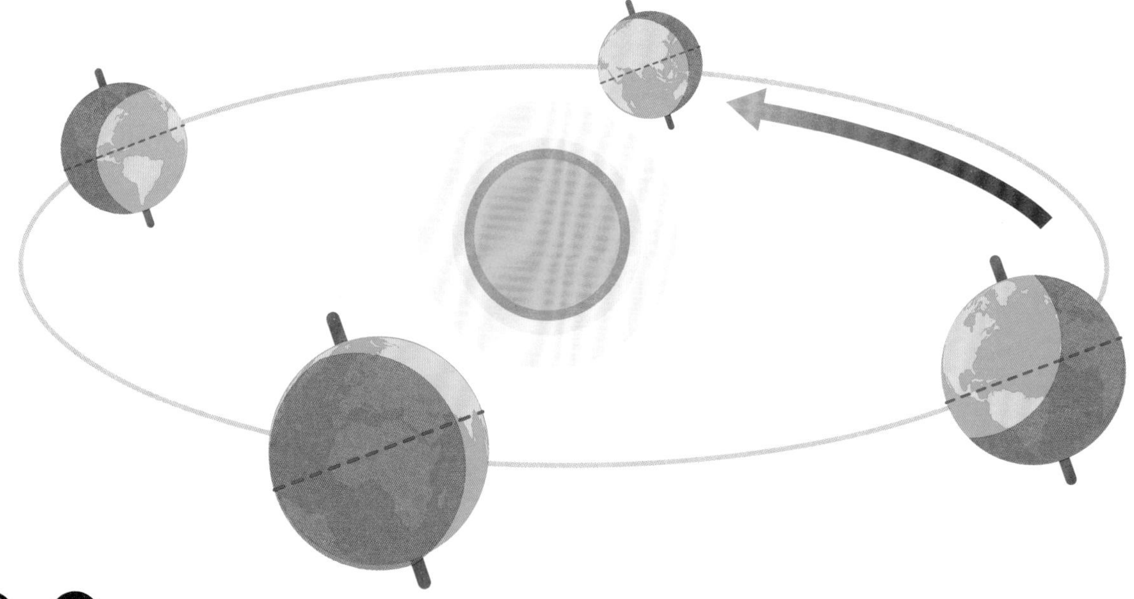

Time to do Activity 22 in the Activity Book!

How God Gives Seasons

If you stood in front of a fire to warm up, you might notice that the front of your face gets hot, but your ears do not. The heat is directly hitting your forehead, but it is hitting your ears at an angle.

The same thing happens to the earth. When the sun's light is directly hitting one place on Earth, that place is hotter than when the sun's light is hitting it at an angle. When the sun directly hits one spot, it is summer in that spot. It is winter when the sunlight is hitting the spot at an angle instead of directly. During the times between summer and winter, it is spring or fall for that place. Summer, fall, winter, and spring are called **seasons**.

23.5°

We have changing seasons on Earth because the earth is tilted.

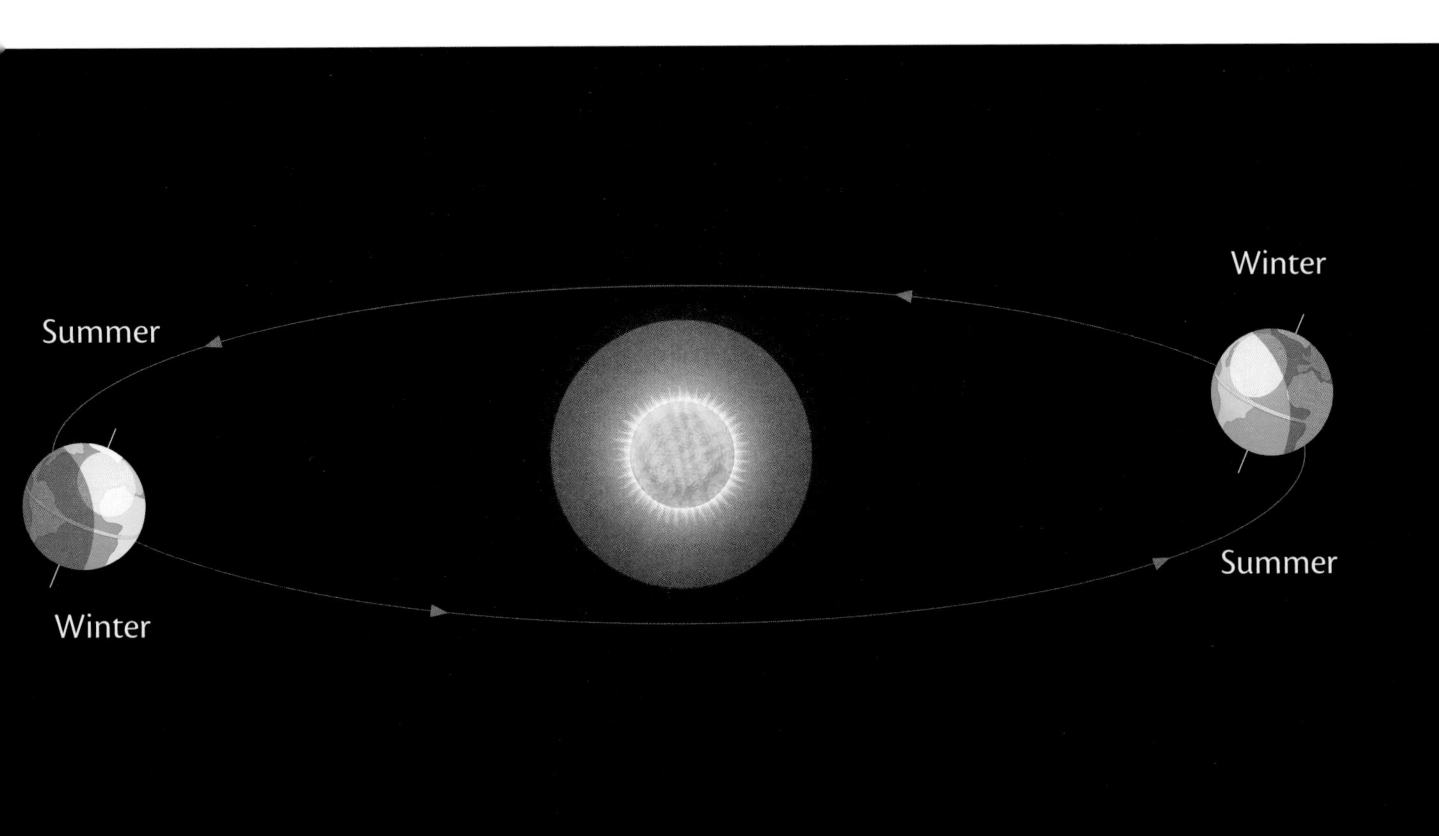

The tilt of the earth always stays the same as it travels around the sun. That means that during part of the year, the bottom half of the earth is tilted towards the sun, and another part of the year, it is tilted away.

When the bottom half is tilted away, the top half is tilted toward the sun. At that time, it would be summer for the top half and winter for the bottom half. Then, once the earth travels halfway around the sun, it would be winter for the top half and summer for the bottom. During the summer, the days are longer than the nights. During the winter, the nights are longer than the days.

Between summer and winter we have fall. Between winter and summer we have spring. Because of the earth's tilt, days and nights are about the same length of time in fall and spring.

Time to do Activity 23 in the Activity Book!

God Protects His Creatures

When you feel cold in the winter, you can put on a coat or go inside to get warm. Wild animals can't do that. They have to stay outside all year long. But God has given them special ways to keep warm during the winter. Here are some ways animals keep warm in wintertime:

1. They go somewhere warmer.

Some birds, whales, and insects travel thousands of miles to find a warmer place to live during the winter. They often eat a lot before they begin their journey. That way, their bodies store up fat to use for energy on the trip. They find their way by using things they see, like stars, the shape of the land, and special light patterns. They also have a special way of feeling direction in their bodies. God has taught them which way to go to find warmer weather. He also teaches them when they should leave.

Gray whales swim about 12,000 miles (19,000 km) each year.

You might think that these animals learn the way by traveling with others who have gone before. But that isn't possible with **monarch butterflies**. The journey is so long that they can't make it all the way in one lifetime. Older butterflies travel to a warm place. Then they lay eggs. The eggs hatch and grow into butterflies at their winter home. These butterflies also have babies there, and those babies have babies. Finally, the butterflies that return to the summer home are the great-grandchildren of the ones that started the trip. How do they know which way to go? God has put something inside them so they know where to go!

Arctic terns fly 44,000 miles (71,000 km) each year.

Monarch butterflies fly about 4,800 miles (7,800 km) each year.

2. Some animals hibernate.

The bodies of furry animals make their own warmth from the food they eat. But in the winter, it may be hard to find food under the snow. God makes the bodies of some furry animals slow down and get cooler so that they don't need much food. Their hearts slow down, and they breathe less. Bears don't even need to empty their urine for months!

> **Definition**
>
> To **hibernate** means to spend the winter sleeping and not moving much.

Animals get ready for hibernation by growing more fur and eating lots of food to get fat. Then they find a sheltered place and go to sleep. These are a few of the furry animals that hibernate:

Bears

Bats

Groundhogs

Hedgehogs

The wood frog has special blood for living through frozen winters.

3. God gave some animals special blood to keep them alive in cold weather.

Frogs have a special kind of blood that doesn't freeze in cold weather. All frogs' bodies slow down whenever it is cold. Some frogs dig a hole to spend the winter in. Others sit on the bottom of a pond where it doesn't freeze. Some even spend winter above ground where it does freeze. But they all have special blood with a lot of sugar in it. The sugar keeps the blood from freezing in the winter.

4. Some animals huddle together.

Penguins, **snakes**, and **bees** gather closely together to share warmth.

O LORD, You preserve man and beast.
(Psalm 36:6)

Bees

Penguin chicks

Rattlesnakes

Prayer

Lord, You are glorious! Your light and warmth shines on the earth every day of the year. You have tilted the earth just right to give it seasons. These changing seasons allow things to grow on earth. They are also interesting and beautiful! We praise You for all the ways You made Your plants and animals, helping them to do well in each season. Amen.

Time to do Activity 24 in the Activity Book!

UNIT 3

Air, Sky, and Space

Hello! I came to visit because we are going to learn about air! I help give breath to living things. I twirl in the sky to make wind and weather. We hope you like this children's hymn!

When Puff is finished with air, I get to show you lights in the heavens! Here is our new memory verse!

Memory Verse

The darkness and the light
are both alike to You.
(Psalm 139:12)

Hymn Singing

Who Made Ocean, Earth, and Sky?

Who made ocean, earth, and sky?
God, our loving Father.
Who made sun and moon on high?
God, our loving Father.
Who made all the birds that fly?
God, our loving Father.

Who made lakes and rivers blue?
God, our loving Father.
Who made snow and rain and dew?
God, our loving Father.
He made little children too;
God, our loving Father.

You can listen to this hymn by searching for "God, Our Loving Father children's hymn" on the internet.

CHAPTER 9
Air

Invisible Air

When you finish eating a bowl of soup, you say the bowl is empty because you don't see anything in it. But the bowl isn't really empty. It is full of air!

Air might look like empty space, but this isn't all it is. There is a lot of empty space in air, but it also has tiny pieces of gas in it that are too small to see. These pieces are spread far apart. Air is a gas.

Remember, gravity pulls on air to keep it close to the earth so that we can breathe. That means air is pushing on us. We don't feel the push because it is pushing on us from every direction, even from inside us. We do feel air pushing when we are out in the wind.

As air, Puff has many important jobs to do. She likes doing them, and she does them well.

Puff's List of Air's Jobs

1. To blow around the earth, keeping places from getting too hot or too cold.
2. To blow clouds around so they can water God's earth.
3. To bring oxygen inside of people and animals when they breathe in.
4. To take carbon dioxide out of people and animals when they breathe out.
5. To bring carbon dioxide into tiny holes in the leaves of plants.
6. To take oxygen out of tiny holes in the leaves of plants.
7. To give plants nitrogen.
8. To protect us from the heat and cold of outer space.

"The Spirit of God has made me,
And the breath of the Almighty gives me life." (Job 33:4)

Time to do Activity 25 in the Activity Book!

What Is Air Made Of?

Air has different kinds of gas in it. One of these gases is called **oxygen**. This is the gas your body uses each time you take a breath. Oxygen is used for everything your body does. You breathe it in. Then your blood carries it through your body to the different workplaces that use it. You could not live without oxygen.

God put the perfect amount of oxygen on Earth for people and animals. If there was too little oxygen, our bodies would not work. If there was too much, our bodies would speed up inside and wear out quickly.

Another gas in air is **nitrogen**. Nitrogen helps spread out the oxygen so we don't get too much in one breath. It helps in other ways too. Fire needs oxygen to burn. If air was all oxygen, fires would start too easily, spread quickly, and be hard to put out. Nitrogen helps keep this from happening.

God put tiny creatures in dirt. Some of these creatures are called bacteria. We can't see these creatures, but they can take nitrogen out of the air and give it to the plants. When animals or people eat the plants, their bodies use the nitrogen to make muscles.

Carbon dioxide is another gas found in the air. This is what plants need to make food for themselves.

Time to do Activity 26 in the Activity Book!

Flying

People have always been amazed at how birds fly. In the Bible, we can read a list of things in God's creation that are hard to understand. One of those things is the way an eagle flies (Prov. 30:19).

People have learned that hot air rises. We can use hot air to rise too. We have been able to rise up into the sky in hot air balloons. People can even travel in them by using the wind.

Birds can do better than hot air balloons! God made the wings of birds in a special way. These wings can use His air perfectly to fly up and glide down, to

Air is heated
inside a balloon
to make it rise.

speed up and slow down, to flap and twist when changing direction, and to stay still while soaring. Now we can travel in airplanes because people have learned about bird wings. The shape of a bird's wing makes the air that goes over the top of the wing travel faster than the air going underneath. This makes the wing lift up and causes the bird (or plane) to fly!

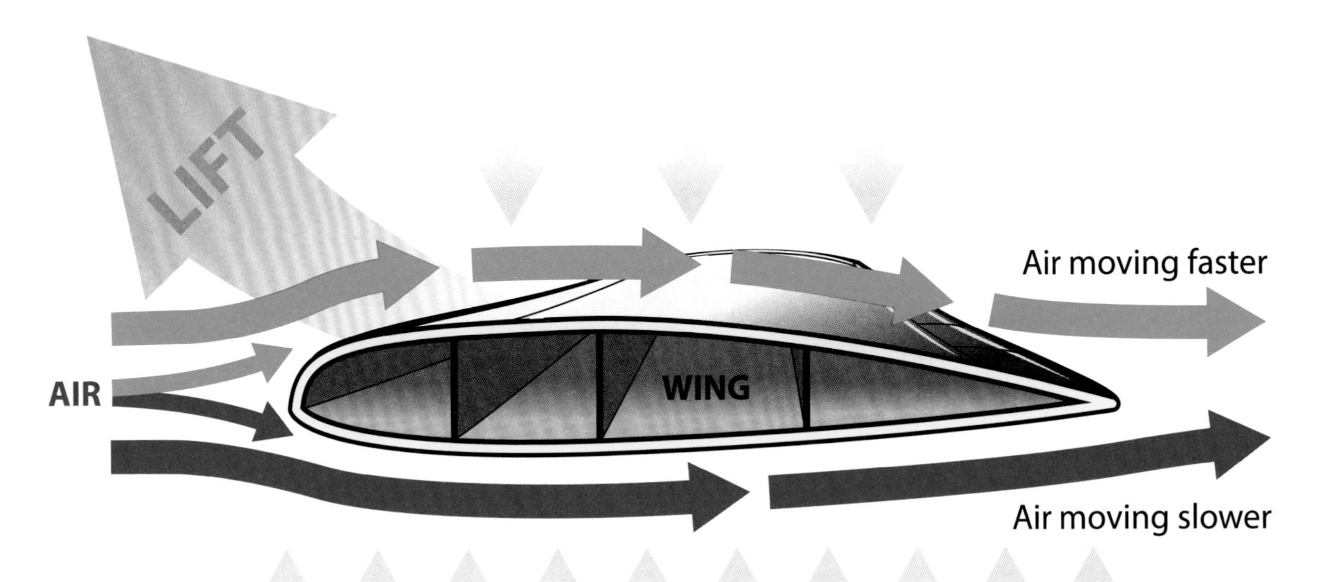

Prayer

Thank You, Father, for the right kind of air to breathe: oxygen for us and carbon dioxide for plants. Thank You for nitrogen that travels from the air to the dirt to help plants become healthy food for us. Thank You that birds can fly. We praise You for giving people wisdom to learn about birds so they can invent machines that can fly. Amen.

Time to do Activity 27 in the Activity Book!

CHAPTER 10
Sky

On a sunny day, the sky is blue. Do you know why? The sky is blue because gases up there are reflecting only blue light for us to see.

Layers of Sky

God made the earth special so that there could be living things on it. That's why He made the sky on the second day of creation. It protects Earth from the heat and cold of outer space. The sky has five layers with different jobs:

1. The first layer, closest to the earth, has gases important for life. It is also where weather happens.
2. The second layer has a different gas that protects Earth from dangerous energy coming from the sun.
3. The third layer is where space

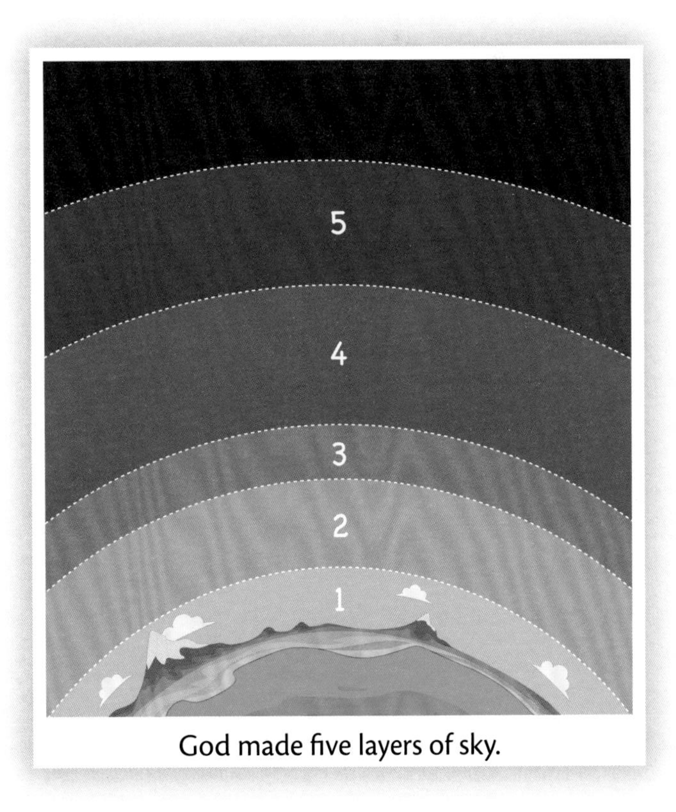

God made five layers of sky.

Northern lights

rocks usually burn out before they can hurt the earth.

4. The fourth layer is where the beautiful northern and southern lights are. This layer also protects Earth from more dangerous energy coming from the sun.
5. The fifth layer is almost outer space, but it still has a little gas that Earth's gravity keeps in.

Time to do Activity 28 in the Activity Book!

Water from the Sky

We have learned that **clouds** are part of the water cycle. Clouds are made of water vapor. This vapor has risen up from the ground all the way to the first layer of sky. We don't see this vapor floating up because the pieces of water are so small.

Definition

A **cloud** is a group of tiny drops of water or ice that can be seen in the air.

The sky gets colder the higher you go. As the vapor goes higher, the cold air makes the small pieces of water join together and become bigger pieces. If enough pieces gather together, we can see them as clouds.

If the drops keep getting bigger, they will become heavy enough to fall down as rain.

Rain falling from clouds.

"Have you entered the treasury of snow . . . ?" (Job 38:22)

Snow is a treasure! Each little flake has amazing beauty. A backyard covered with bright, clean snow calls children out to play. Snow reminds us that God can clean us from the bad things we do. He says that even if our sins are like yucky red stains, He can make them as white as snow.

Snowflakes come in many beautiful designs.

Snow usually falls as flakes. Snowflakes are flat and have six points. They are always shaped that way because the very tiniest piece of water always has a certain shape and must join with other tiny pieces in a certain way. If the air is cold enough, each flake starts forming around a piece of dust in the sky. First comes a six-sided shape. Then points are built on, and beautiful decorations grow on each point. No one has ever seen two snowflakes that are exactly alike. God would know if there has ever been any!

Hailstones are another kind of frozen water that falls from clouds. In the Bible, snow is a blessing from God, but hail is not. Sometimes God used hail to punish or chase away His people's enemies.

Hail is made differently than snow. In summer, when air near the hot earth quickly rises, water in the clouds is blown higher to cooler air where it freezes. The frozen drops start to fall, but they are blown up again by the wind, and more water freezes on them. This keeps happening until the hailstones are too heavy to stay in the sky any longer. Then they fall down to the ground.

Time to do Activity 29 in the Activity Book!

Weather from the Sky

When He utters His voice—
There is a multitude of waters in the heavens:
"He causes the vapors to ascend from the ends of the earth;
He makes lightnings for the rain;
He brings the wind out of His treasuries."
(Jeremiah 51:16)

Definition

Weather is what God sends us from the sky. Sunshine, wind, and water falling from clouds are all part of weather.

When the sun shines on the earth, the earth gets warm. The warm earth heats the air above it. Hot air does not stay there. If it did, Earth would get too hot for us. Instead, the air goes up! When the hot air rises, new air comes in to take its place. This air is cooler and gets sucked in as the hot air is leaving. We can feel this new air coming in. If it comes softly, it is a breeze. If it comes quickly, it is wind. It's nice to have dirty air blown away and fresh air brought in!

If the hot air going up has a lot of water vapor in it, and the air it goes into is dry and cold, a very tall cloud can be made. Then, lots of wind blows all different speeds and directions in the cloud. These winds give the cloud energy. The ground also has energy. These two energies want to meet. If the cloud gets low enough, and if there is a high place for the ground's energy to climb up, they reach toward each other and make a ZAP! of lightning. Lightning is very bright and very hot. It is five times hotter than the sun!

If you see a tornado, go inside to a basement or inner room.

When lightning zaps, it makes a loud sound called thunder. A storm with a lot of thunder and lightning is called a **thunderstorm**. There is a lot of rain and wind in a thunderstorm.

Sometimes thunderstorms make a dangerous wind called a **tornado**. Fast wind twists around and around between a cloud and the ground. Tornadoes can blow leaves off the trees and a house off its foundation.

Prayer

Thank you, God, for all the things Your sky gives us: protection from rocks and dangerous energy coming from space, rain and snow to give water, and wind for fresh air. Please protect people and houses where tornadoes sometimes happen. Amen.

Time to do Activity 30 in the Activity Book!

CHAPTER 11

The Moon: Our Nightlight

The moon is nighttime comfort from God. It does not make its own light like the sun does, but it reflects the sun's light. It helps us see in the dark, but it isn't so bright that we can't sleep. Wherever we go on Earth, we can still see the moon. It's nice to go on a trip away from home and see our same moon in a new place!

"His throne . . . shall be established forever like the moon,
Even like the faithful witness in the sky." (Psalm 89:36-37)

Besides giving the moon for a nightlight, God has given it other important jobs. It helps make the earth special so people, animals, and plants can live on it.

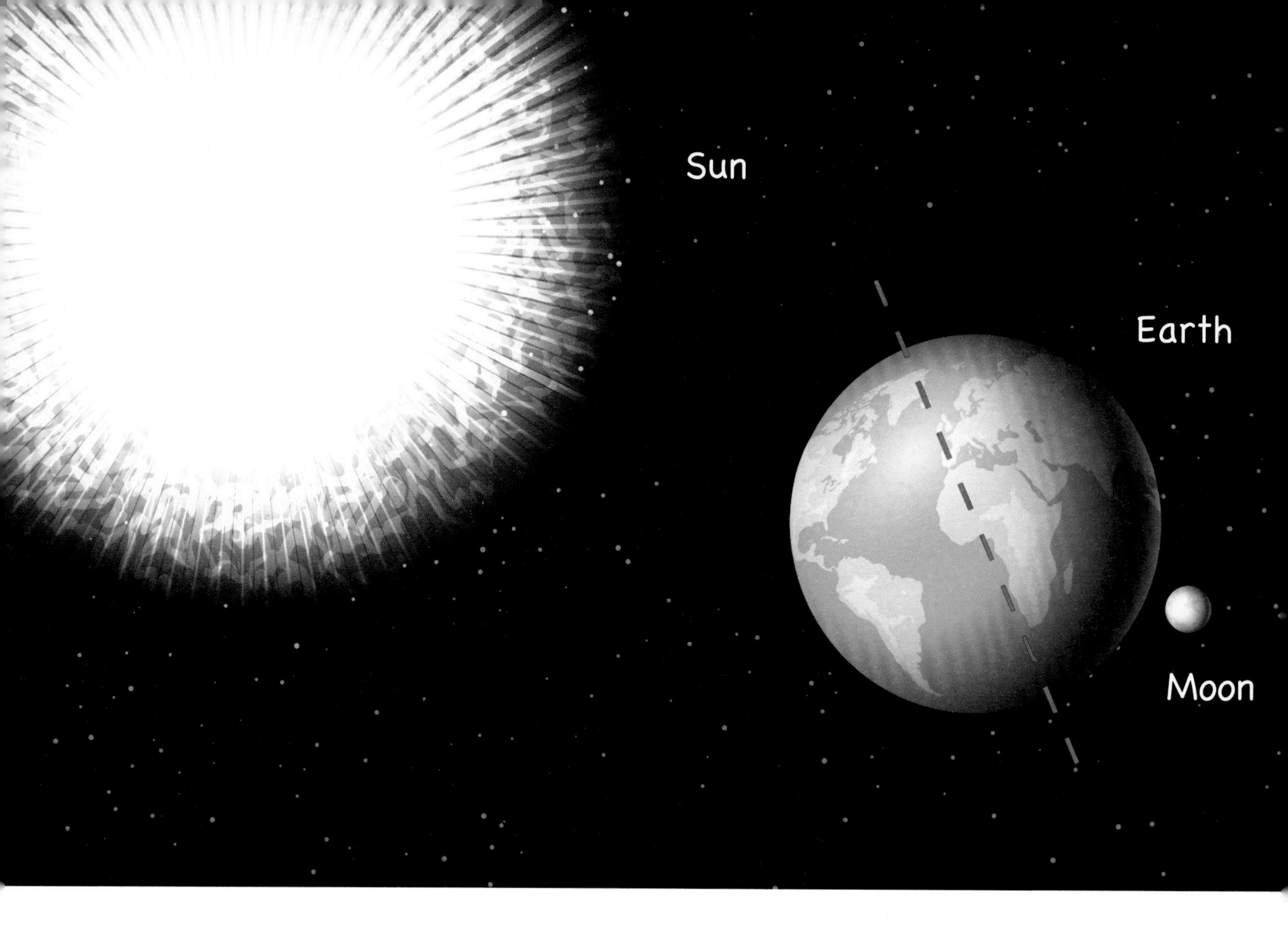

The Moon Keeps the Earth Tilted

While Earth is traveling around the sun once a year, the moon is traveling around the earth about once a month. Do you remember that the earth is tilted as it goes around the sun? Well, the moon seems to pretend that the earth is not tilted. The moon goes around the earth as though it is paying more attention to the sun than the earth. It does not go around Earth's equator!

God made the moon big enough that there would be a lot of gravity pulling between it and the earth. This pull, and the funny way the moon travels around it, keeps Earth tilted. The tilt is the reason we have seasons. The seasons keep Earth's temperature friendly for life in most places. If Earth were not tilted, a lot of the earth would be desert, and a lot of it would be covered in ice.

Time to do Activity 31 in the Activity Book!

The Moon Keeps Seashores Clean and Alive

The moon's gravity pulls on the earth. It also pulls on the ocean's water. We call this a **tide**. When a certain part of the ocean is turned closest to the moon, its waters pile up because the moon is pulling on them. This makes the water deeper. When this happens, this part of the ocean is having high tide. At high tide, the waves come farther up beaches and higher on sea cliffs.

Tides are changes in the ocean caused by the moon's gravity, making each part of the ocean first deeper, then shallower.

Because the water gets deeper at that place, it must get shallower somewhere else. That shallower place is having low tide. During low tide, the water does not come far up the beach. Rocks that were once under water become dry.

As the earth spins, the moon is pulling on the ocean at different places all the time. This is why we have high and low tides all over the world. Seashores have two high tides and two low tides every day as the earth spins.

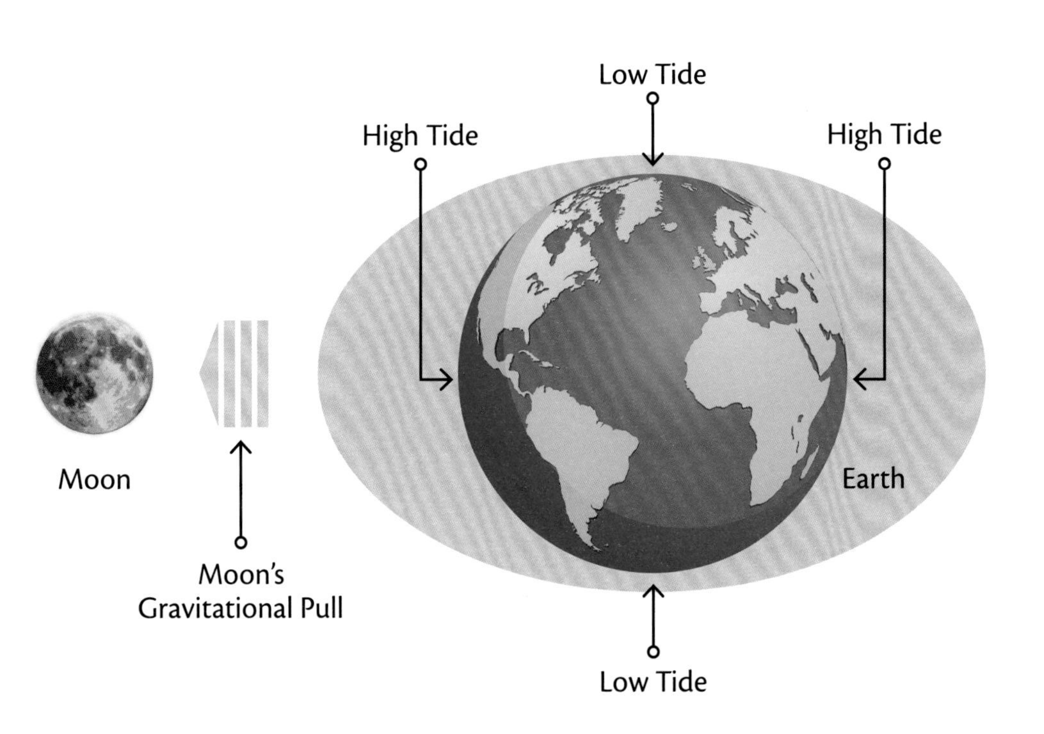

Because of the moon's pull, the ocean moves a lot at its shores. These tides and waves help clean the shore. The moving water also helps little creatures that live near the shore. It brings food to these creatures. But moving water can be dangerous. Because of crashing waves, these little sea creatures must hang on to the rocks so they don't get washed away. They have special ways of grabbing food as it comes by them underwater.

Girl looking in tide pool

God has made many different kinds of creatures to live in this area that is always moving and changing. When the tide is low, it is fun to look for little pools of water left behind on the shore and around the rocks on the beach. These tide pools have interesting creatures that have been left by the waves and other creatures that have stayed attached to the rocks.

During a total solar eclipse, the moon completely covers the sun.

The Moon Helps Us Learn about the Sun

If the sun is shining in your eyes, you can put your hand up to block it. Even though the sun is very big and your hand is little, this works because your hand is close to you.

Definition

A **solar eclipse** happens when the moon comes between the earth and the sun and blocks part or all of the sunlight coming to Earth.

Sometimes the moon comes between the earth and the sun just like your hand does. We call this a **solar eclipse**. Usually only part of the sun is blocked. But sometimes the moon completely covers the sun. When this happens, people can learn things about the edge of the sun without hurting their eyes. God is so good to put the moon just far enough away to perfectly cover the sun.

Time to do Activity 32 in the Activity Book!

Let's Get To Know the Moon!

- The earth has more iron on it than anything else, but the moon has no iron on it.
- The moon has less gravity than Earth because it is smaller. We could not breathe on the moon because there is not enough gravity to keep air there.
- Like the earth, the moon spins in a counterclockwise direction.
- Like the earth traveling around the sun, the moon travels around the earth in a counterclockwise direction.

- Because of how fast the moon spins and because of how fast it travels around the earth, we always see the same side of the moon. We never see the other side.
- When you look up at the moon, sometimes it looks like a round ball. Other times it might look like a crescent or half of a pie. The moon might seem to change shape during the month, but it doesn't really change. It always stays the same, but it looks different to us because it is moving. As the moon travels around us each month, we see the sun shine on it in different ways. When the moon is full (a bright circle), we are seeing the sun shining straight on it. When we see only part of the moon lit, we are seeing the sun shining sideways on it. When the sun is shining on the other side of the moon, we don't see the moon at all. We call this a new moon.

Prayer

Father, thank You for giving us the friendly moon to see at at night. You put it in just the right place to do its job for the earth, the oceans, and all of the life You placed here. Amen.

Time to do Activity 33 in the Activity Book!

Milky Way as seen from Earth

CHAPTER 12

Outer Space

God made the lights in space on the fourth day of creation. When we see these big, wonderful things that God made, it reminds us how small we are. We are amazed that, even though we are so little, God thinks of us and sent Jesus to us. In the Bible we read that David felt the same way:

**When I consider Your heavens, the work of Your fingers,
The moon and the stars, which You have ordained,
What is man that You are mindful of him,
And the son of man that You visit him?
(Psalm 8:3-4)**

Our Solar System

Our earth is not the only thing traveling around the sun. There are seven more planets in space traveling around it. There are also comets, asteroids, meteoroids, dwarf planets, and moons. These things are part of what we call the **solar system**.

Definition

Our **solar system** is made of the sun (a star) and everything that travels around it.

Comet Hale-Bopp, Arizona, U.S., 1997

A **comet** is a large chunk of ice with rocks and dust mixed in. In its path around the sun, a comet might spend years far away from us in the solar system and only a few months close by. When it comes close to the sun, the sun melts some of the ice off the comet. This melted ice becomes vapor. Then the vapor is blown off the comet away from the sun, making a tail millions of miles long.

A **meteoroid** is a piece of rock in space that probably broke off a comet. If it comes into our sky, it starts burning. Then it is called a **meteor** or a shooting star. If it doesn't burn up and we find it on the ground, it is called a **meteorite**.

An **asteroid** is a rock that travels around the sun. Asteroids can be many shapes and sizes, but they are too small to be called planets. Most of our solar system's asteroids are in a belt that is between the four smallest planets and the four largest.

Long ago, a meteorite made this crater in Arizona, U.S.

Time to do Activity 34 in the Activity Book!

Planets

Definition

A **planet** is a large object, made by God, that travels around a star.

Planets reflect the sun's light just like the moon does, but they are so far away from us that they look like stars. It is only with a telescope that we can see their round shape.

Here are the four small planets closest to the sun. They are all made of rock. They are listed in order, starting with the one nearest the sun.

1. Mercury is the littlest planet in our solar system. It has no air around it, so it gets very hot where the sun shines, and very cold where it does not.

2. Venus is almost as big as Earth. It has air around it that would be poisonous for us to breathe. Its air is so thick that it holds onto heat. Venus is the hottest planet.

3. Earth is the only planet specially made by God for people, animals, and plants to live on.
4. Mars is about half the size of Earth. It has a lot of rusty iron in its dirt, so it looks reddish. There are huge red dust storms on Mars.

Here are the four largest planets. They are all made of gas.

5. Jupiter is our biggest planet and has about 70 moons. It has a big red spot. This spot is a storm that is always blowing, but it isn't always in the same place.
6. Saturn is smaller than Jupiter but is still very big. Around it are thousands of rings made of rocks, ice, and dust. It has about 80 moons.
7. Uranus is smaller than Saturn, but it's still very big. Instead of spinning like an ice skater as all the other planets do, it spins sideways, like a wheel. It has about 20 moons.
8. Neptune is about the same size as Uranus. Its gas makes it look blue. Neptune has 14 moons.

A **dwarf planet** is larger than an asteroid but smaller than a planet. A planet's moon cannot be a dwarf planet. There is one dwarf planet in the asteroid belt and four more past Neptune. Pluto is the dwarf planet closest to Neptune.

Time to do Activity 35 in the Activity Book!

Outside Our Solar System

Our sun is a star, close and friendly. It is a medium-sized star. At night, we see other stars far away. Some are very large. Large stars are the hottest stars and look blueish. Some stars are smaller than the sun. Small stars burn cooler and give off reddish light.

He counts the number of the stars;
He calls them all by name.
Great is our Lord, and mighty in power. (Psalm 147:4-5)

God made galaxies in different shapes.

Definition

A **galaxy** is a big group of millions or billions of stars, gas, and dust held together by gravity.

The **universe** is everything that is anywhere, and all the space between.

There are more stars in the sky than anyone could count. But God can, and He even has names for them all! Guess what: Stars burn to make their own light!

We are in a **galaxy** called the **Milky Way**. Why is it called this? If you go outside at night, far away from street lights and other light, you can look up and see a whitish streak across the sky. This whitish streak may look like milk, but it is actually a lot of stars that we are seeing as we look through a thick part of our galaxy. This is why it's called the Milky Way.

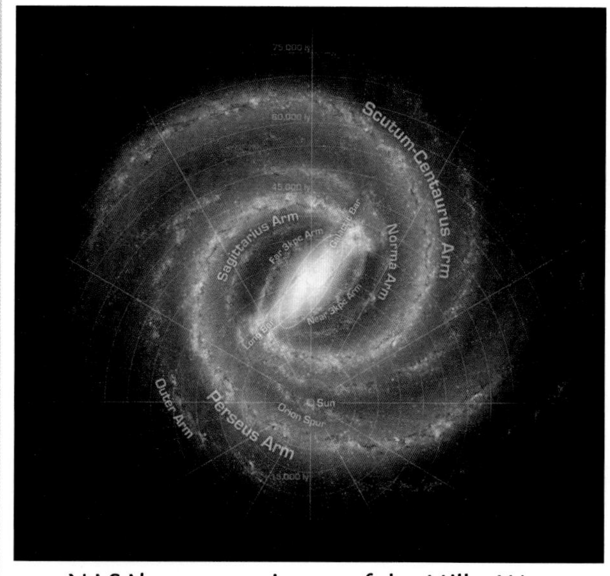

NASA's current picture of the Milky Way

Our galaxy isn't the only one in the sky. God made billions of galaxies! Outer space has more galaxies than anyone could count. God made many different shapes for these galaxies. The Milky Way has five arms swirled into a flat spiral. It is spinning like all galaxies are thought to do. Other galaxies have different shapes. These galaxies are all part of the **universe**.

Only God knows if the universe stops or goes on forever. Only He can understand such amazing things!

Nebula: a cloud of gas and dust in outer space

Prayer

Lord, You are so big! When we look up, we see such wonderful, faraway things that You made. Thank You that You made little children too! Amen.

Time to do Activity 36 in the Activity Book!

UNIT 4
Plants

It will be so fun to learn about plants with you! Plants have an interesting way of using light (me) to make food for themselves. That's good because plants can't move around or go places to get their food!

Here is our Unit 4 memory verse! It is a song (psalm) about fields being joyful because God is coming!
Our hymn is a carol that sings about joyful fields when Jesus came!

Memory Verse

Let the field be joyful, and all that is in it.

Then all the trees of the woods will rejoice before the LORD.

For He is coming.

(Psalm 96:12-13)

Hymn Singing

Joy to the World

Joy to the world! the Lord is come:
Let earth receive her King;
Let every heart prepare Him room,
And heav'n and nature sing,
And heav'n and nature sing,
And heav'n, and heav'n and nature sing.

Joy to the earth! the Savior reigns:
Let men their songs employ;
While fields and floods, rocks, hills, and plains
Repeat the sounding joy,
Repeat the sounding joy,
Repeat, repeat the sounding joy.

You can listen to this carol by searching for "Joy to the World for kids" on the internet.

CHAPTER 13

The Parts of a Plant

God made plants on the third day of creation. Three days later He made the first two people, Adam and Eve. He gave them the job of taking care of a beautiful garden called Eden, and He told them why He had made plants:

And God said, "See, I have given you every herb that yields seed which is on the face of all the earth, and every tree whose fruit yields seed; to you it shall be for food." (Genesis 1:29)

Let's Learn about the Parts of a Plant:

Roots hold the plant in the ground. They take water and minerals in from the dirt. They often store food for the plant (or hungry people and animals) to use later.

Stems help the plant stand up so its flowers and leaves can do their jobs. Stems also take water upward in the plant and carry food downward to the lower parts of the plant.

Leaves take in sunlight and carbon dioxide to make food for the plant. They put out oxygen for people and animals to breathe.

Flowers make pollen. Without pollen, there would be no seeds. God made some flowers big, pretty, and sweet with nectar so they would attract the insects

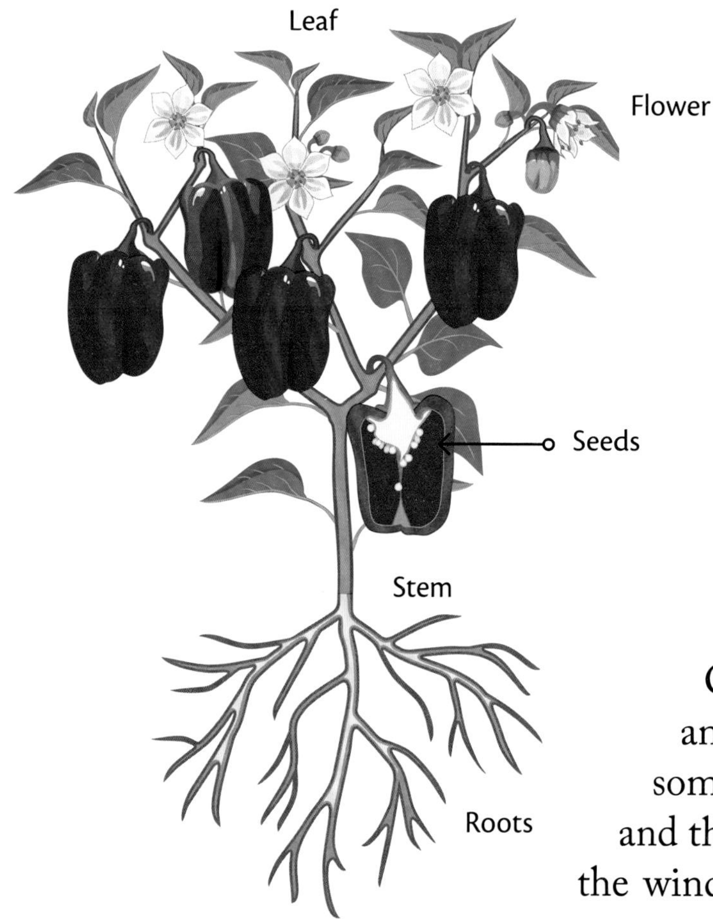

needed to carry their pollen. Other flowers that are small and hard to notice get their pollen carried around by the wind.

Fruits grow after the flowers have done their job. Fruits shelter the seeds until they are grown up. God made fruits as a way for plants to spread their seeds somewhere else so they can grow new plants. Usually, the fruits taste good. Animals eat them, but the seeds pass through their bodies and come out somewhere else. Other seeds are in fruits that stick to an animal's fur (or your socks) and come off somewhere else. Some fruits can burst open and throw their seeds, while others can float in the wind. Only God could make plants with so many ways to spread all over the earth.

Seeds have a little food inside them and all the instructions they need to become another plant. The new plant will be the same kind as the one it came from.

Time to do Activity 37 in the Activity Book!

Liquid Moves Inside Plants

Without water, plants cannot make the food they need. They need water to stay alive. They must get water from the dirt their roots are in. God has made an interesting way for water to move upward into plants even though gravity is pulling it downward. He made water stick easily to itself and to other things. If water is in a small enough tube, its stickiness is stronger than gravity and makes the water move up.

The stems of plants have small enough tubes for this to happen. Some of the tubes inside the stems take water and minerals upward. Even though one of these going-up tubes is only about as thick as a human hair, it moves a lot of water. The many tubes in a corn plant put out about 53 gallons (200 L) of water into the air during a summer. A large rainforest tree can put out about 300 gallons (1200 L) a day![12] If more water goes out of the plant than what comes in, it will wilt.

There is another kind of small tube inside stems that takes liquid food made by the leaves down to the other parts of the plant. These tubes have a way to keep liquid from rushing straight down. That way, food has time to spread all through the plant.

Time to do Activity 38 in the Activity Book!

Plants as Food

- **Flowers**: Before flowers bloom, they are called buds. Some flower buds we eat are artichokes, broccoli, cauliflower, and cloves (a spice)

- **Fruits**: Apples, cherries, pineapples, mangos, tomatoes, peppers, cucumbers, squash, pumpkins, berries, grapes, olives, black pepper, lemons, oranges, avocados, chiles, paprika,

Freshly dug potatoes and carrots

cranberries, and melons

- **Seeds**: Nuts, corn, peas, lentils, pinto beans, rice, wheat, cumin (a spice), sesame seeds, sunflower seeds, oats, barley, and coffee
- **Leaves**: Lettuce, spinach, kale, cabbage, tea, and herbs (oregano, basil, thyme, sage, parsley, cilantro)
- **Stems**: Celery, asparagus, rhubarb, bamboo, sugarcane, cinnamon (bark from tree trunks), maple syrup (sap from tree trunks); and onions, potatoes, and ginger, which are all underground stems
- **Roots**: Sweet potatoes, carrots, beets, radishes, turnips, and jicama

Prayer

Thank You, God, for the delicious things You gave us to eat from plants. They are so colorful and yummy, and they keep us healthy. Amen.

Time to do Activity 39 in the Activity Book!

CHAPTER 14

The Life Cycle of Plants

A Plant Begins

When Jesus first chose people to follow Him, He only picked a few of them. But Jesus wanted His friends to understand that soon more and more people would believe in Him, so He told them this story:

Then He said, "To what shall we liken the kingdom of God? . . . It is like a mustard seed which, when it is sown on the ground, is smaller than all the seeds on earth; but when it is sown, it grows up and becomes greater than all herbs, and shoots out large branches, so that the birds of the air may nest under its shade." (Mark 4:30-32)

Seeds start out small. But inside them they have all the instructions they need to become a whole plant or tree. Seeds have a dry, hard cover to protect them until they can be planted in warm, watered dirt. Inside

this hard cover, the seed has the beginnings of a tiny plant and one or two big pieces of food for the tiny plant to use. This food feeds the new plant until its roots are in the dirt and its leaves are in the sunshine.

In warm, watered dirt, the seed swells until its cover splits open. God has made plants able to feel gravity. The root comes out of the seed and grows down into the earth. A stem comes out and grows up. This stem has the big pieces of food attached to it. The new plant has been using those pieces of food so far. They are its first leaves, and they soon start making more food for the plant. They do not look like the leaves that will grow later, and they will fall off when the true leaves take over the food-making job.

Time to do Activity 40 in the Activity Book!

A Plant Grows

As the plant grows, its stems get taller and fatter. God made stems able to lean and curve to help the plant get the most light. They do this by stretching the shaded side of the stem. Sunflowers can stretch different sides of their stems all day long to keep their young flowers facing the

Definition

Pollen is a powder traded between flowers. Each tiny piece has some seed-making instructions in it.

sun. Then, when the flowers get older and need **pollen** to make seeds, they only face the sunrise. God made sunflowers this way to attract chilly insects that are looking for a warm flower's nectar and will bring along some pollen.

Time to do Activity 41 in the Activity Book!

Pollen + Eggs = Seeds

God gave living things a way to make more of their kind. Animals and people can make more of their own kinds by having babies. Plants make more of their own kinds by making seeds.

Seed-making starts with flowers. Most flowers have petals. Flowers also have pollen-making parts and pollen-taking parts. After the petals wither and fall off, the flower makes a fruit for the seeds to grow in.

Petals are the parts of flowers that look and smell so nice to us. They are also the parts that God made to attract the insects and birds that carry pollen.

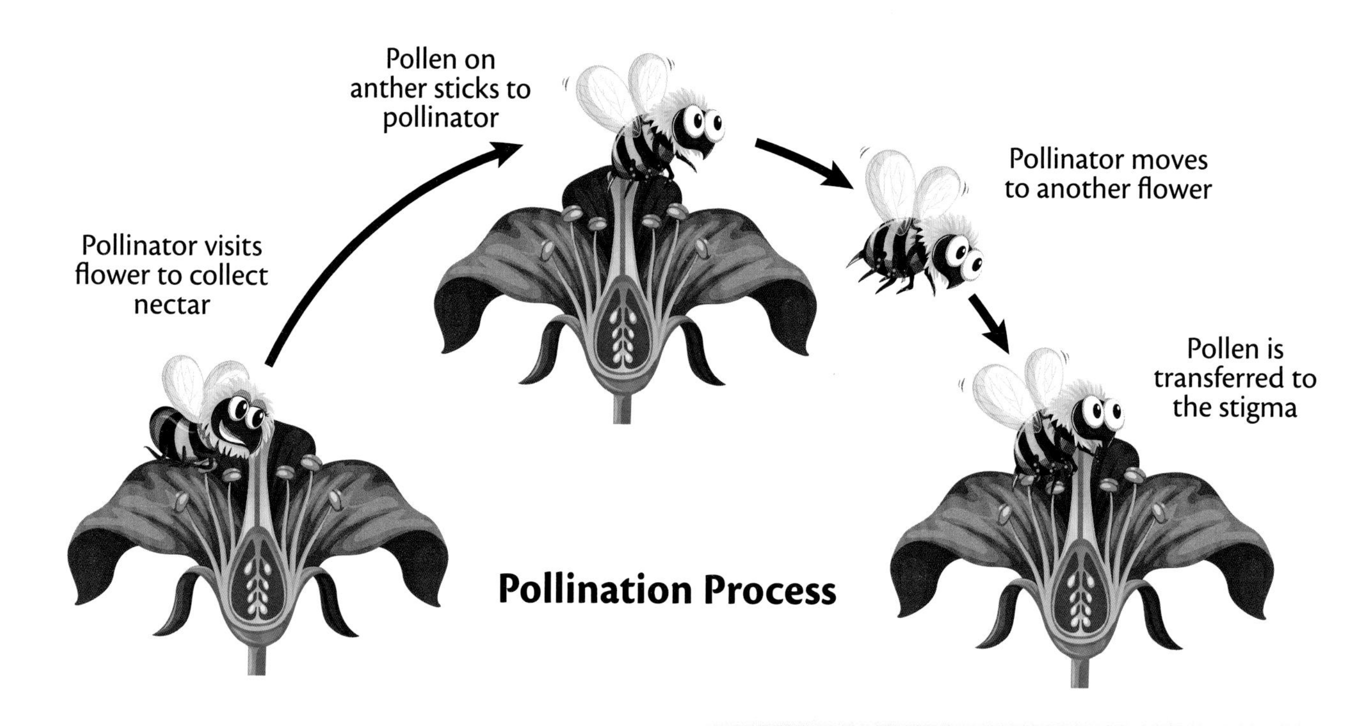

Definition

Pollinators visit flowers, carrying pollen from flower to flower as they look for nectar to eat.

Nectar is the sweet liquid made by plants that attracts pollinators.

Anthers are the pollen-making parts of a flower.

Bee pollinating flower

Animals that carry pollen are called **pollinators**. These pollinators come to the flower because they think it has something sweet for them to eat. This sweet thing is called **nectar**.

When a pollinator sees or smells the petals of a flower, it comes to check for nectar. As it crawls into the flower, it brushes against the **anthers** where pollen is made. Some of the pollen sticks to the pollinator.

Flowers don't use the nectar they make. God gave flowers this nectar because it is what the pollinators need, even though nectar doesn't matter to the plant. This way, the flowers help the pollinators.

When bees go out to look for flowers, God made them to carry pollen to the same kind of flower during each trip from the hive. This is the amazing way God made sure that flowers would get the pollen they need from their own kind of plant. They can't use pollen from a different kind of plant. The bees don't care what kind of flower they visit, but God knows that this is what the plants need. Flowers and bees don't know that they are helping each other, but God does!

Each piece of pollen has only half of the instructions it needs to make a seed. The other half is in an egg. A plant has to put pollen and an egg together before it can make a seed. A plant's eggs are inside the part that will become a fruit. When a pollinator goes from one flower to the next, it takes pollen with it. Each flower has a sticky place near the eggs

Definition

Stigmas are the pollen-taking parts of flowers. Their tops are sticky or hairy to help them grab pollen.

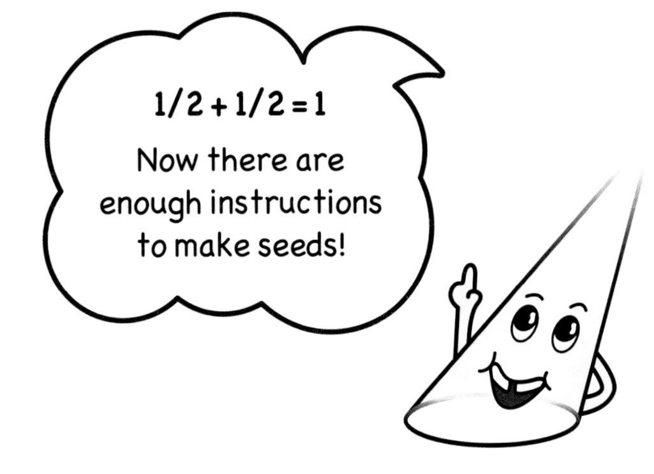

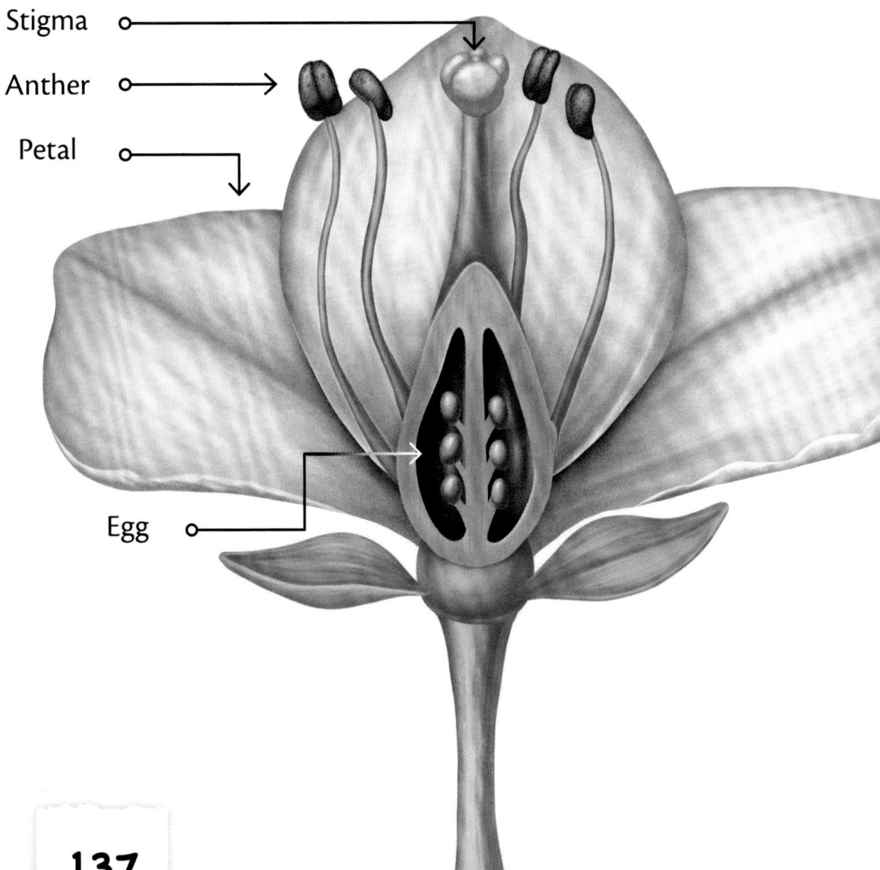

that pollen will stick to. This sticky place is called a **stigma**.

Once a piece of pollen gets stuck to the stigma, the pollen opens up. A long tube comes out and takes the pollen's half of the seed-making instructions down to where the eggs are.

Next, the fruit, with the seeds inside, starts to grow. The seeds are growing too. Do you remember that a cycle is when things happen in a certain order and end up where they started? Well, when the plant's seeds are all grown up, they are ready to be planted and begin the plant's life cycle again!

Some plants die each year after making seeds. They fall over, rot, and become part of the soil. As part of the soil, they help keep it moist and give important nutrition to the next plants that grow there. Other plants, like trees, can live and make seeds for many years.

Prayer

We praise You, God, for the way plants and insects work together as You planned. Thank You for making seeds that become plants and make more seeds that become plants. Amen.

Time to do Activity 42 in the Activity Book!

Water lilies

CHAPTER 15

How Plants Make Food

When Jesus didn't want His friends to worry about where they would get their food and clothes from, He told them to think about a flower: Even though a flower doesn't work hard like we do, it grows and is beautiful because God the Father cares for it.

"Consider the lilies, how they grow: they neither toil nor spin; and yet I say to you, even Solomon in all his glory was not arrayed like one of these." (Luke 12:27)

Little Green Machines

God gave all plants a way to grow by making their own food. The way they do this is too small for us to watch, but people have learned about it by using microscopes. Microscopes make small things look bigger. With a microscope, we can see that leaves have a lot of tiny machines that make more than enough food for the plant.

Dad and son using a microscope

These machines live mostly on the tops of leaves because they need energy from sunlight to do their job.

Leaves are green because these tiny machines are green. Inside each machine is a lot of green stuff that speeds up the work of making food. This green stuff is called **chlorophyll**.

Definition

Chlorophyll is the green stuff inside plants that speeds up their work of making food.

God has given plants chlorophyll so they will make more food than they need. The extra food is stored in the plant and can be food for people and animals to eat.

Time to do Activity 43 in the Activity Book!

Photosynthesis

On the underside of leaves, we find a different kind of tiny machine. Each one looks like a little mouth with lips that open and close. We call them **stomata**.

Definition

Stomata are the tiny openings on plants that allow gases to go in and out.

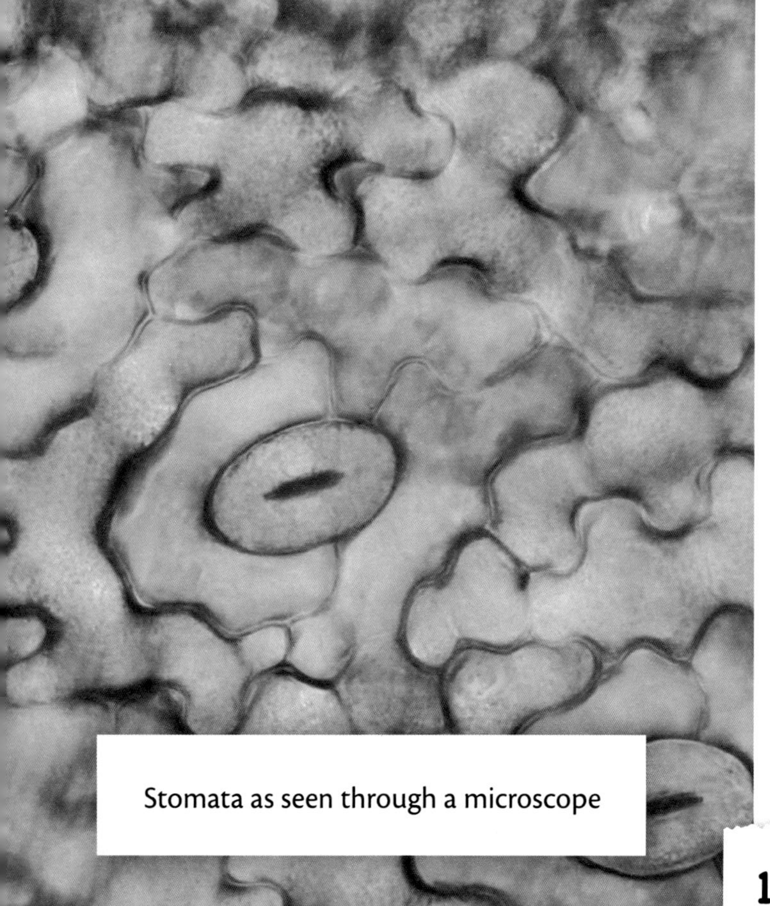

Stomata as seen through a microscope

The green machines on the top of plant leaves need carbon dioxide to make their food. Carbon dioxide is a gas. The stomata open up so they can get that gas out of the air. While the stomata are open, they also let out oxygen into the air. Oxygen is a gas the green machines make while they work, but they don't need this gas, so they send it out into the air. People and animals need to breathe in that oxygen. Then, when people and animals breathe out, carbon dioxide leaves their bodies as waste. Plants need that carbon dioxide to make food for themselves and for us. People and plants work together to help each other. How perfectly God made it!

Now that we know about the tiny machines, we can learn about how food is made in **photosynthesis**!

> **Definition**
>
> **Photosynthesis** is a plant's way of making food.

There are four things that the green machines use to make food. I am one of them. My three friends are water, carbon dioxide, and chlorophyll. Without all of us, plants could never make food!

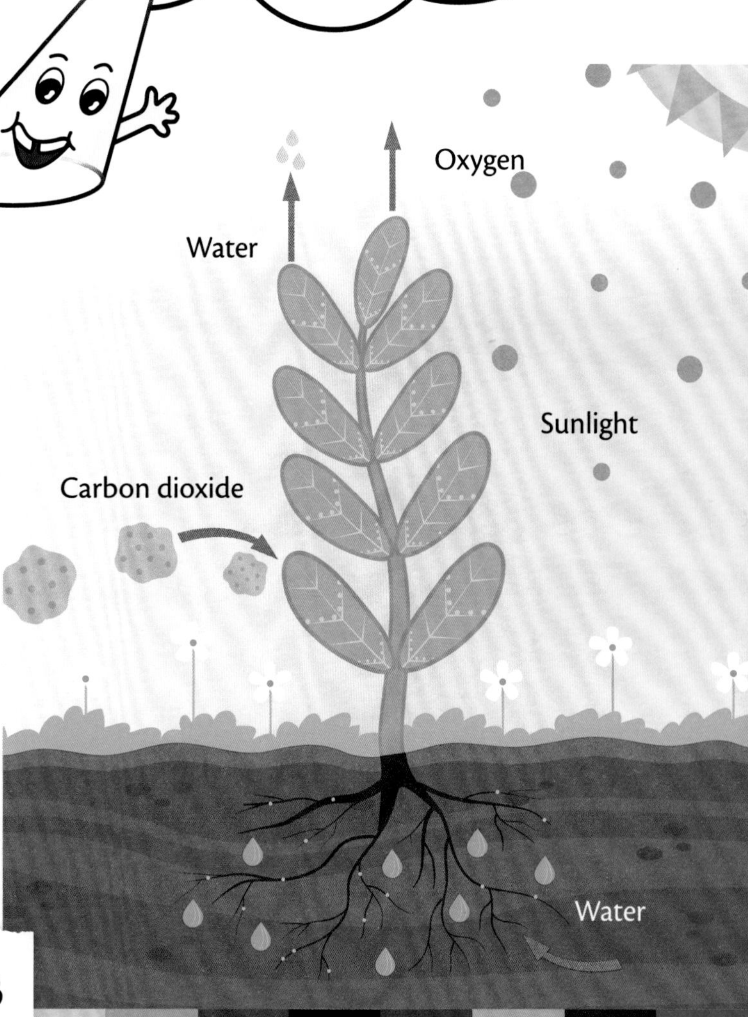

In Photosynthesis:

1. Water travels from the plant's roots to its leaves.
2. Carbon dioxide comes into the leaves from the air through the stomata.

3. The green machines make food out of the water and carbon dioxide. They use sunlight for energy. The food they make stays in the plant to help it grow and make seeds. Some food is stored in its roots.
4. Oxygen goes out into the air from the stomata.

Time to do Activity 44 in the Activity Book!

Falling Leaves

God has made a way for **deciduous** trees to save water in winter. Since so much water is lost through leaves, the trees drop them! What makes these leaves fall off?

As days get shorter, something happens to the place where the leaf's little stem attaches to the branch. It hardens and blocks the liquids that usually go up and down through the little tubes to the leaf. Without water, the green machines cannot make food. The chlorophyll in the leaf gets used up and fades away. Sometimes there are other colors we can't see in leaves because the green is hiding them during the summer. When the green is gone in the fall, we see beautiful yellows, oranges, or reds. If there are no other colors hidden by the green, the leaves just turn brown. Liquid from the

Definition

Deciduous trees are trees that lose their leaves in the fall.

Fallen leaves are fun to play in!

tubes that carry things down and up is sent to the roots for the winter. Then, because the leaves have no water, they die and fall off. Deciduous trees take a break from photosynthesis!

Some trees do not lose their leaves in the fall.

In places near the equator, where days have the same amount of light all year long, leaves do not change color and drop.

Evergreen trees like cedars, pines, and spruces have thin leaves covered in wax that do not lose water easily. Their leaves stay on the tree all winter and do not freeze because of a special liquid in them. They also do not change color.

Prayer

Thank You, God, for photosynthesis! We praise You that plants make what we need to breathe, and that we make what they need to make food. Thank You that they make more food than they need, which You give to us! Amen.

Time to do Activity 45 in the Activity Book!

CHAPTER 16

God's Amazing Variety of Plants

Since no book could tell about all the plants God has made, let's learn about different groups of plants instead. The Bible tells us that God made two groups of plants: **trees** and **herbs**.

Trees Are Woody Plants

**The righteous shall flourish like a palm tree,
He shall grow like a cedar in Lebanon. (Psalm 92:12)**

God made trees to flourish or grow so He could give us their wood as a very sturdy gift. Wood is so sturdy that when a woody plant dies, its stems take a very long time to fall, rot, and become part of the dirt. Because wood is strong and tough, it is good for building homes and furniture. We can also burn it to cook our food and keep us warm. The manger that the baby Jesus slept in may have been made of wood.

Shrubs

Tree

There are two kinds of woody plants:

1. **Shrubs** or bushes are shorter than trees. A single shrub has several woody stems growing out of the ground.
2. **Trees** have only one stem growing out of the ground. We call this stem a trunk.

There are two kinds of trees:

Trees that grow their seeds inside a fruit. Deciduous trees grow their seeds inside fruits. Their wood is harder than the wood from other kinds of trees.

Trees that do not grow their seeds inside a fruit. Conifers are one kind of tree whose seeds don't grow in fruits. These trees make **cones** instead of fruits. Wind takes pollen from pollen-making cones to other pollen-taking cones on the tree. Conifer seeds grow in the open air, resting on the layers of the pollen-taking cone. Coniferous wood is softer than wood from deciduous trees.

God has made an interesting way for conifers

to help each other! They work together to protect other trees. If one lodgepole pine tree is attacked by pine beetles that have brought a sickness, it will put out something like a smell into the air. The other lodgepole pines nearby will notice the warning and start making changes to themselves so that pine beetles will not like their taste! This protects them from the harmful beetles.[13]

Some of these lodgepole pines are dying because of pine beetles. Others were able to protect themselves.

Time to do Activity 46 in the Activity Book!

Herbs

**He causes the grass to grow for the cattle,
And vegetation for the service of man,
That he may bring forth food from the earth. (Psalm 104:14)**

When we talk about herbs, we sometimes mean the leaves of plants like oregano or basil that we use to flavor food. But when God said He made **herbs** at creation, He meant all the plants without woody stems. These plants are smaller than trees, and they don't live as long as trees do. Herbs die every year. Some grow again from seeds. Others come back from roots buried in the ground.

People have separated herbs into two groups:

Oat field

1. **Grasses** are usually short plants with long, narrow leaves. Grains such as rice, oats, corn, and wheat come from grasses. Animals with hooves often eat grass. Many grasses grow back from their roots if they are eaten. People who raise grazing animals usually move the animals around to different fields. This gives the grass a chance to grow back. Although grass is usually short, one kind of grass is very tall: bamboo!

2. **Forbs** are short plants with wide leaves and showy flowers. Any herb that is not a grass is a forb. Many forbs are pollinated by insects. We know of one of

Many forbs that bloom in spring grow from onion-like bulbs underground.

God's flowers that makes its nectar sweeter when a bee is buzzing nearby. It seems to almost hear the sound! It only takes three minutes for the nectar to change.[14]

Time to do Activity 47 in the Activity Book!

Surprising Plants

Every plant in God's created world is wonderful! But many plants surprise us because they don't look or act like most other plants. Let's look at two of them.

The **Victoria water lily** from South America grows its roots in mud at the bottom of lakes. Its stems can be 26 feet (8m) long. The stems reach all the way from the bottom of the lake to the surface of the water. At the end of the stems are round, floating leaves which can grow 20 inches (51cm) a day until they are 10 feet (3m) wide. Each leaf has an interesting rim around its edge.

Victoria water lily

The Victoria water lily has a big, beautiful flower that only lives two days—long enough to get pollen. One day, a big bud floats up. The next day, at sunset, it opens, showing a flower so white that it seems to glow as the day darkens. Beetle pollinators soon come because the flower gets warmer than the air around it and gives off a nice smell like pineapple and bananas. By morning there are several pollen-covered beetles inside the flower. Then the flower closes and traps them inside, where they eat and pollinate all day. That night, at sunset, the flower opens again, and out fly the beetles to pollinate somewhere else with this flower's pollen. This flower has now changed its color to pink or red. As daylight comes, it sinks into the water for its seeds to grow.

The **saguaro cactus** is one of the largest cactus plants, growing up to 40 feet (12m) tall. It starts very small, does not get branches for many years, and makes its first flowers at about age fifty! Its flowers are only open for one day each year, but that is enough time to get pollen for making its red fruits.

Cactus plants live in deserts where there is not much rain. Their roots spread over a large area to get as much water as they can when it does rain. Their green stems can store water by getting fatter. Without water, the stems get skinnier and folded. It is not a fun surprise if you touch a cactus! Instead of leaves, they have spines which protect them.

Saguaro cactus in bloom

Prayer

We praise You, God, for water lilies that live in water and for cactus plants that live without it most of the year. The world You gave us is full of amazing trees and herbs! Thank You for causing them all to grow. Amen.

Time to do Activity 48 in the Activity Book!

UNIT 5

Crunchy Creatures

There are millions of kinds of creatures with crunchy skin! Some cause trouble for us; others help. Have you ever watched ants busily doing their work? God uses them to show us how to be diligent.

Our hymn reminds us that God made all creatures—even the small ones!

Memory Verse

Go to the ant, you sluggard!
Consider her ways and be wise, . . .
[She] provides her supplies in the summer,
And gathers her food in the harvest.
(Proverbs 6:6, 8)

Hymn Singing

All Things Bright and Beautiful

Each little flower that opens,
Each little bird that sings,
God made their glowing colors,
He made their tiny wings.

Chorus:
All things bright and beautiful,
All creatures great and small,
All things wise and wonderful,
The Lord God made them all.

The cold wind in the winter,
The pleasant summer sun,
The ripe fruits in the garden—
He made them, every one.

You can listen to a cheerful tune for this hymn by searching for "All Things Bright and Beautiful — Church Hymn" on the internet.

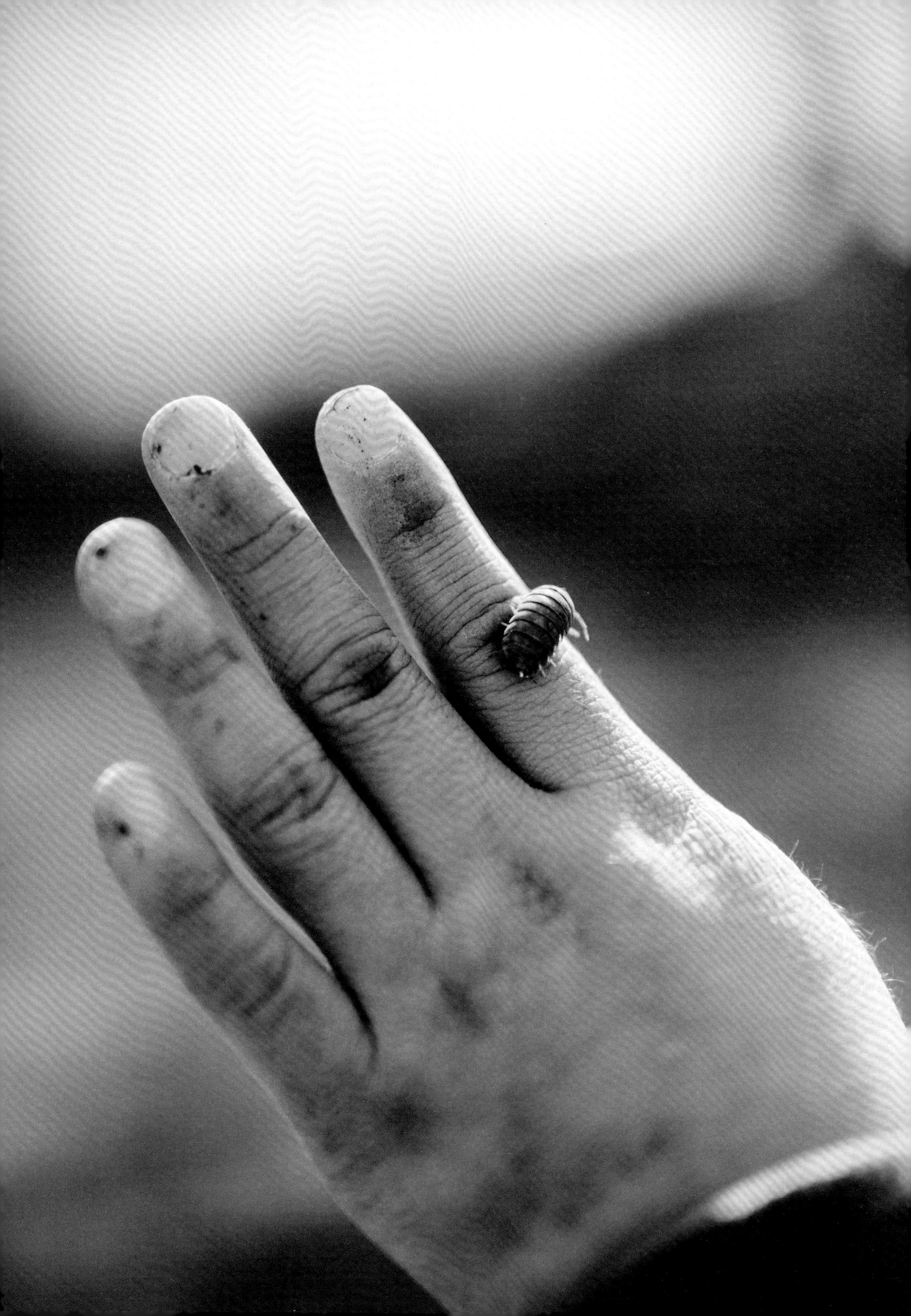

CHAPTER 17

What Makes Bugs Crunchy?

When you go outside to catch bugs, you are usually looking for insects and spiders or maybe millipedes and centipedes. You might also find pill (or roly-poly) bugs.

Crunchy Skin

All these creatures are crunchy because they have hard skin. These bugs have no bones to help them hold their shape. Instead, their hard skin makes them sturdy and able to stand up. This skin acts the same way our bones do. Our muscles work by attaching to our bones, but their muscles attach to the inside of their skin. This skin is called an **exoskeleton**. All insects have exoskeletons. Other creatures like crabs have them too.

> **Definition**
>
> An **exoskeleton** is the hard skin on the outside of some creatures.

Mexican redknee tarantula shedding its skin

The exoskeletons of crunchy creatures cannot stretch and grow like your skin does. As these animals grow, a new exoskeleton forms underneath

Small red crab

the old one. The one on top gets weaker and one day splits open and falls off. The new exoskeleton is larger than the old one so that the animal inside can grow. Often, the new skin has new parts. It might have wings for an adult insect. It might even have a new leg for a crab who has lost one to a hungry bird.

These creatures also have crunchy antennae coming out of their heads. They use their antennae to feel, smell, and taste things. Their legs, their mouth parts, and their eyes are also parts of their exoskeletons. A bee's clear wings or a butterfly's colorful ones are pieces of exoskeleton too.

Besides having hard skin, all crunchy creatures:

- Have babies that hatch from eggs.
- Do not breathe the way you do. You bring air in and out through your nose and mouth. These creatures have openings on different parts of their bodies where air passes through their exoskeleton.
- Cannot warm their own bodies. They are always just as hot or as cold as the air around them. Their activity slows down when they are cold, and it speeds up when they are hot. Because these creatures can't warm their own bodies, we call them **cold-blooded**.

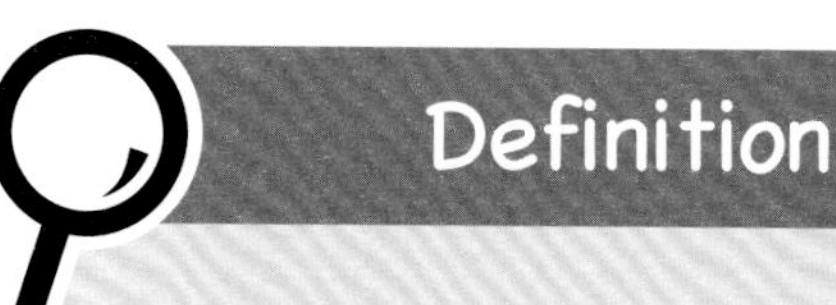

Definition

Cold-blooded animals are not able to warm their own bodies. They are the same temperature as the air or water around them.

Time to do Activity 49 in the Activity Book!

Springing, Singing, and Flinging

Grasshoppers can jump very far. If they were as big as a person, they could jump across a football field! Their back legs have a special knob of exoskeleton inside each knee. When a grasshopper is resting, it slowly bends its legs until they are folded tightly. A muscle in each knee gets stretched around that knob. When it's time to jump, the muscle relaxes enough to suddenly slip off the knob and the grasshopper springs into the air! Its jump works the same way your finger and thumb do when you snap your fingers or flick a bug off your sleeve.

Crickets make nice, summertime chirping noises. Their exoskeleton at the edge of each wing is made differently. One wing has a long row of tiny, tooth-like bumps. It looks like one side of a zipper. The other wing has something to scrape along the row of teeth. Cricket chirping works the same way as if you scrape your thumb nail along the teeth of a comb. Cricket songs are very loud because of the special way the wings are shaped.

Time to do Activity 50 in the Activity Book!

Silver-spotted skipper butterfly: its caterpillar can shoot frass 6 feet (2m) away.

Caterpillars seem to be always eating. Because so much goes in their mouths, a lot comes out the other end. The waste that comes out of insects is called frass. All that frass can be a problem for caterpillars, because its smell lets their wasp enemies know where they are hiding. God has made some caterpillars with an exoskeleton machine that can shoot their frass far away. Before the piece of frass comes out, it is loaded onto the machine inside the caterpillar. The machine backs up until it clicks into a catch that holds it. Then the caterpillar uses its muscles to push. When it pushes hard enough, the catch lets go, and flings out the frass!

Silver-spotted skipper butterfly larva

Prayer

Thank You, God, for the creatures with exoskeletons. Their bodies work very differently than ours, but You made them perfectly for what they need to do. Even though they are amazing animals, I'm glad my skin is soft and my bones are inside me instead of on the outside. Amen.

Time to do Activity 51 in the Activity Book!

Larger cabbage white caterpillars are pests of cabbage in Europe.

Dragonfly

CHAPTER 18

What Makes an Insect Special?

A man named John was sent by God to get people ready for Jesus to come. John lived outside in God's creation, away from people. But people went to hear him and be baptized. In the verse below, can you see two ways that John used insects?

Now John himself was clothed in camel's hair, with a leather belt around his waist; and his food was locusts and wild honey. (Matthew 3:4)

How Are Insects Different from Other Crunchy Creatures?

Insects have:

1. Three body sections so they can move their stiff exoskeletons.
2. Six legs attached to the middle section.
3. Wings (usually four) attached to the middle section.

4. Holes in their exoskeletons for breathing. Small insects can move enough air through these holes without doing anything. Large insects need to open and close certain holes while squeezing their tummies to push air through their bodies so they can breathe.

A bee is an insect because it has three body sections.

Definition

Metamorphosis is the surprising change in how some animals look and act as they grow from eggs to adults.

5. Changes that happen to their bodies when growing from eggs to adults. This change is called **metamorphosis**.

Time to do Activity 52 in the Activity Book!

Metamorphosis

Some insects hatch out of their eggs and look just like their adult parents except that they don't have any wings. This kind of insect will shed its exoskeleton a few times as it grows.

After shedding one last time, it will have wings. Getting wings all of a sudden would be a big change! The insect is now an adult and can have babies.

Other insects, like bees and butterflies, go through a much bigger change! When they hatch out of their eggs, they don't look like their parents at all. They look more like worms. This type of baby insect is called a **larva**.

A larva eats and eats, and grows and grows, shedding its skin several times. The last time it sheds, it is not a larva anymore. It is a hard, round case or shell with no mouth, eyes, or legs. It is a **pupa**. A pupa of a butterfly or moth is called a chrysalis. Before a moth becomes a chrysalis, it makes something around itself called a cocoon. A butterfly does not make a cocoon before becoming a chrysalis.

Inside the pupa's hard shell, the young insect's insides first turn into liquid. There are no muscles and no stomach—nothing but liquid and tiny pieces that will one day become the adult insect's body parts. These tiny pieces were formed inside the larva before it even hatched from its egg! God is the One who made it all work that way. It is too amazing for us to

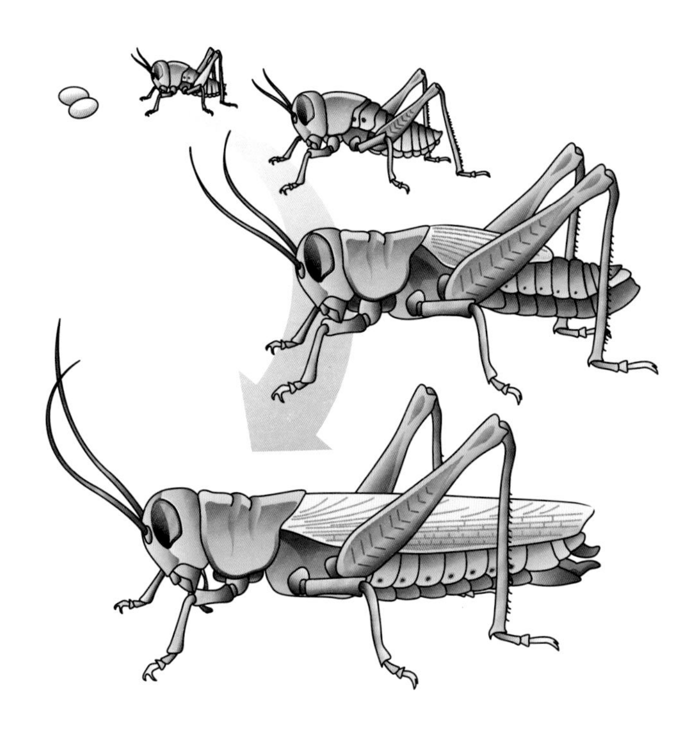

Grasshopper Metamorphosis

Definition

A **larva** is a baby insect that does not look at all like its parents but looks more like a worm.

Definition

A **pupa** is an insect in its hard shell when it is changing from a larva to an adult.

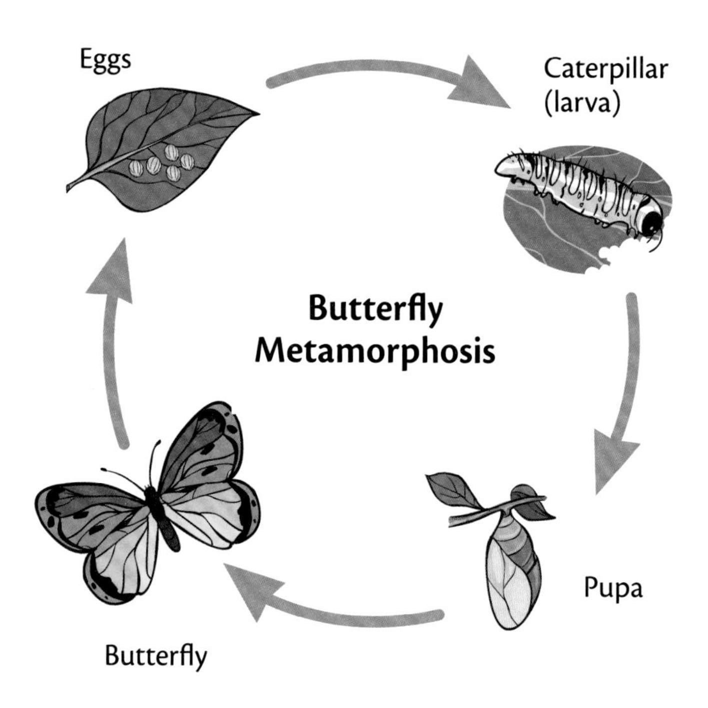

understand!

Then, as two weeks or so go by, wings, legs, and all the things needed to make an adult insect are formed inside the pupa. At just the right time, the adult struggles to open its case and crawls out. At first, its body and wings are soft and easily damaged, but in time they harden. The larva has been completely changed! Now it has three hard body parts, six legs, wings, antennae, and all it needs to have babies.

God made such a wonderful, mysterious surprise with metamorphosis. He has a wonderful, mysterious change for us too. This change will happen when Jesus comes back to Earth. Our bodies will be raised from the dead. They will be new bodies and we will live forever with Jesus.

Behold, I tell you a mystery: We shall not all sleep, but we shall all be changed—in a moment, in the twinkling of an eye, at the last trumpet. For the trumpet will sound, and the dead will be raised incorruptible, and we shall be changed. (1 Corinthians 15:51-52)

Time to do Activity 53 in the Activity Book!

Working with bees to get honey

Helpful Insects

[God's Word is] sweeter also than honey and the honeycomb. (Psalm 19:10)

Insects are helpful to us in many ways. Here are a few of those ways:

1. Many insects help with pollinating flowers, but bees do most of this job. We would not have as much food or as many different kinds of food without insects.

2. Bees make their homes out of wax from their bodies. People can use this beeswax in candles, lotions, and lip balm.

3. Honey is made from flower nectar that bees have brought back to their hive. In the hive, the bees fan it with their wings to make it thicker and sweeter. Then we get to enjoy eating it!

Silkworm and cocoon

4. One kind of food coloring (called Red 4) is made from certain South American insects that live on cactuses. The insects are squished, and their red juice is saved. Red 4 is healthier than man-made red food coloring.

5. Silkworms are caterpillars that make cocoons when it's time to become a chrysalis. They make their cocoons by squirting out a liquid that becomes a thin, strong thread that they wind around themselves. Long ago, people learned how to unwind these cocoons and weave the thread into the cloth we call silk.

6. Insects eat up messes! Some flies and beetles eat dead animals and animal waste. Some ants, beetles, and termites eat dead trees. Many insects eat dead herbs. As insects eat, their frass adds good things to the dirt to help plants grow well.

7. Some insects help us by eating other insects that would ruin the food we grow.

8. Insects are food for many other animals.

Ladybug eating a harmful aphid

Prayer

Lord, we don't understand how metamorphosis works. It seems like You make beautiful butterflies out of nothing, just like You did when You created all things. Thank You for reminding us of Your creation miracles! Amen.

Time to do Activity 54 in the Activity Book!

Lizard eating an insect

CHAPTER 19

Different Kinds of Insects

Insects and other crunchy creatures are some of the creeping things God made on the sixth day of creation.

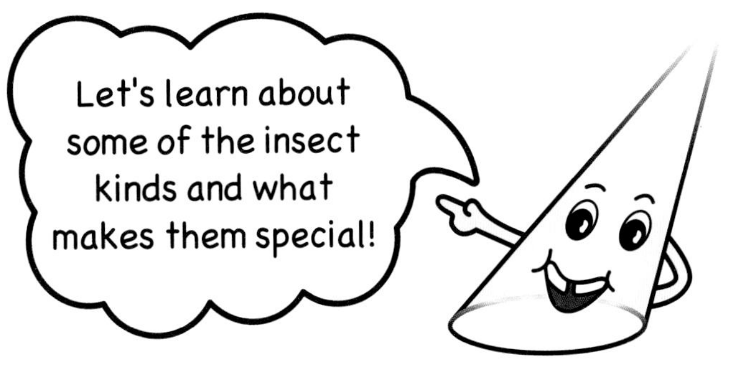

Then God said, "Let the earth bring forth the living creature according to its kind: cattle and creeping thing and beast of the earth, each according to its kind"; and it was so. (Genesis 1:24)

Insects Whose Babies Look Like Their Parents

Dragonflies—Some baby dragonflies spend three years underwater as they grow up. God has given them a special way to get oxygen out of the water. They have gills like fish! They eat tadpoles, little fish, and worms. Then they climb

Dragonfly

Water strider, praying mantis, grasshopper

onto land to shed their skin one last time and become adults.

True Bugs—This group of insects has mouth parts that can poke into a plant or animal and suck juices out. Some kinds of true bugs are assassin bugs, cicadas, squash bugs, aphids, and water striders. Water striders can walk on water with special feet God gave them.

Praying Mantises—With big front legs ending in claws and spines, mantises are ready to snatch other crunchy creatures to eat. They may look like they are praying, but they are really holding up their front legs so they are ready to pounce. Most animals cannot hear the noises bats make, but some mantises that fly at night can. When mantises hear bats making their insect-hunting sounds, they quickly dive and turn circles to get away from the bats.

Grasshoppers—When grasshoppers have enough food to eat, they like to live alone. But when they run out of food, something changes. Grasshoppers eat plants, but if it doesn't rain for a long time, the plants they eat don't grow. Then the grasshoppers get very hungry! When it finally does rain, the plants start to grow again. Lots of hungry grasshoppers find those first plants that come up. With so many grasshoppers coming together, something happens in their brains to make them want to stay together. It makes their bodies change color and their wings grow larger. They lay a lot more eggs more often. A lot of young grasshoppers hatch from the eggs. Then they swarm in big groups and crawl along, eating every plant they come to. The adults, with their large wings, fly far away together, looking for more food. When this happens to grasshoppers, we call them **locusts**. When these grasshoppers start eating our crops, we call it a **plague**.

Time to do Activity 55 in the Activity Book!

Insects Whose Babies Look Like Worms

Ladybug, larva, and eggs

Beetle larvae

Beetles—Some beetles you might know are ladybugs (ladybird beetles), Japanese beetles, dung beetles, weevils, and fireflies. Adult ladybugs help gardeners by eating about twenty aphids a day. Aphids are bugs that damage the plants in our gardens. Big baby ladybugs help even more by eating about 200 aphids a day! Japanese beetles are beautiful, but they damage many plants. Weevils have their mouths and antennae at the end of a funny, long snout.

Beetles have hard, thick top wings that cover their whole body. Their bottom wings are thin and clear. These clear wings stay folded until it's time to fly. After flight, the beetle's muscles pull them in and fold them in two directions. Some beetles' top wings bend light to make beautiful colors. Long ago, people wore colorful beetles as jewelry! One kind of desert beetle has top wings that are specially designed by God to collect drinking water from dew.

Japanese beetle

Wild bee colony

Bees, Wasps, and Ants—Bees live in big groups called **colonies**. A colony has three kinds of bees: a queen, whose job it is to lay eggs; some **male** drones, whose job it is to help the queen; and a lot of **female** worker bees, whose job it is to collect nectar, make honey, and raise baby bees.

Some kinds of wasps live in colonies. In the spring, one queen will start a little nest. She lays eggs in it that hatch into females. Those females build a big nest for their queen to lay more eggs in. They also catch caterpillars

Wasps

Definition

Female means an animal or person that is a girl or woman.

Male means an animal or person that is a boy or man.

for everyone to eat. There is a lot of fighting between the queen and other females that want to be queen. Sometimes the queen will leave with some males to start a new colony. In places where the weather gets cold, a queen finds a sheltered place to spend the winter. The other wasps all die from the cold.

Some kinds of wasps live alone. Some of these mother wasps catch caterpillars to feed their babies in a little nest. Other mothers lay their eggs right inside a caterpillar. The babies hatch and eat the caterpillar from the inside!

Ant hills are made of dirt the ants have brought up while digging their tunnels.

Ants live together in underground colonies. They have a queen, some males, and many female workers. The workers can have different jobs: baby care, tunnel digging, food gathering, or protecting the nest. When the colony gets crowded, the queen and some of the males grow wings. They fly to a new place and start a new colony. Then their wings fall off.

Time to do Activity 56 in the Activity Book!

Flies and Mosquitoes—Most insects have four wings, but these annoying bugs only have two wings. Where the other two wings usually grow, they have little knobs sticking out of their sides. These knobs help them as they fly. With these

Special knobs on this crane fly help it quickly change its steering.

special knobs, these insects can quickly change their steering. The knobs also help flies and mosquitoes keep their heads steady so they can see when they spin. Without these knobs, they cannot fly!

Butterflies and Moths—To us, the wings of a butterfly or moth seem dusty. That dust is actually tiny, flat scales that lie like shingles on a roof. These scales help protect them from danger. Butterflies and moths hardly ever get caught in a sticky spider web. If they fly into a web, their scales easily snap off and are left behind in the web while the butterfly or moth flies free.

Prayer

Thank You, God, for bright and beautiful butterflies, the flight of flies, busy ants and bees, and all the other amazing insects You made. Amen.

Time to do Activity 57 in the Activity Book!

The wing scales of the blue morpho butterfly bounce back only blue light to give it shimmering blue wings.

Dungeness crab

CHAPTER 20

Yummy and Yucky Crunchy Creatures

The Yummy Ones

People think some crunchy creatures are delicious! They catch **lobsters**, **crabs**, **shrimp**, and **crayfish** to cook them for a special treat. These creatures have big claws on their front legs for catching their own food or for pinching someone who is trying to catch them!

Lobster

Most animals with exoskeletons need to get oxygen from the air. But these yummy creatures live underwater. God gave them a special way to get oxygen from the water they live in. He gave them a special body part called **gills**.

> **Definition**
>
> **Gills** are special openings that God made in some animals so they can get oxygen from water.

As water passes over the many feathery parts of a gill, the animal pulls in oxygen. Then it lets out carbon dioxide waste. These crunchy creatures are able to store water in their gills so they

Roly-poly bugs

can spend a little time on land. But they must get back in the water before the oxygen in their stored water is used up.

Roly-poly bugs are a surprise God made. They have exoskeletons and gills but they would die if they were put underwater. They live on land, but need to get water for their gills from moist air. That's why they are usually found under rotting plants.

Time to do Activity 58 in the Activity Book!

Many spiders make orb webs like this one.

Spiders

Some people think spiders are yucky. It's not fun to walk into a spider web, and some spiders have poisonous bites! But we should thank God for spiders because they catch and eat so many insects that bother us. A spider can eat hundreds of very small flies in one day. Some spiders pounce to

Spider catching a fly

catch and eat insects. Others make webs to trap their food.

Spiderwebs are made of strong, bendy thread. Some webs are built in the air to catch flying insects. Other webs are made on the ground to catch crawling insects. Once caught, the insect is bitten by the spider and given a poison that makes it unable to move. Then the spider squirts something in the insect to make its insides become liquid. After a while, the spider can suck out the liquid for a meal.

Spiders have eight legs, only two body parts, and no antennae. They do not have holes in their exoskeletons to breathe through. They do not have gills. Instead, they breathe through book lungs that look like a lot of pages.

A spider does not get caught in its own web. God made spiders very intelligent. When a spider makes a web, it only makes some of the threads sticky. Then it only walks on the threads that aren't sticky. God gave spiders a special non-stick coating on their toes. This way, if a spider accidentally steps on a sticky thread, it can get free easily. It also has springy hairs on its feet that will push the web off. Some spiders eat their webs and use the food energy from them to make new webs the next day.

Time to do Activity 59 in the Activity Book!

Pesky and Painful Crunchy Creatures

God's people were once kept as slaves in Egypt. God sent Egypt ten really big problems, one after another, so the Egyptians would want to let God's people go. These problems were called plagues. Three of those plagues were insects. God sent locusts to eat the Egyptians' crops. This verse talks about the other two problems:

[God] spoke, and there came swarms of flies,
And lice in all their territory. (Psalm 105:31)

We don't often see plagues of insects come to bite us or destroy our crops, but we are often bothered by one or two yucky creeping things. Let's learn about some creatures that we should stay away from. These creatures might seem very scary. But remember that God cares for you and wants you to trust Him and not be afraid.

Creatures that bite or sting to protect themselves:

- Bees
- Wasps
- Ants
- Spiders
- Scorpions

Scorpion

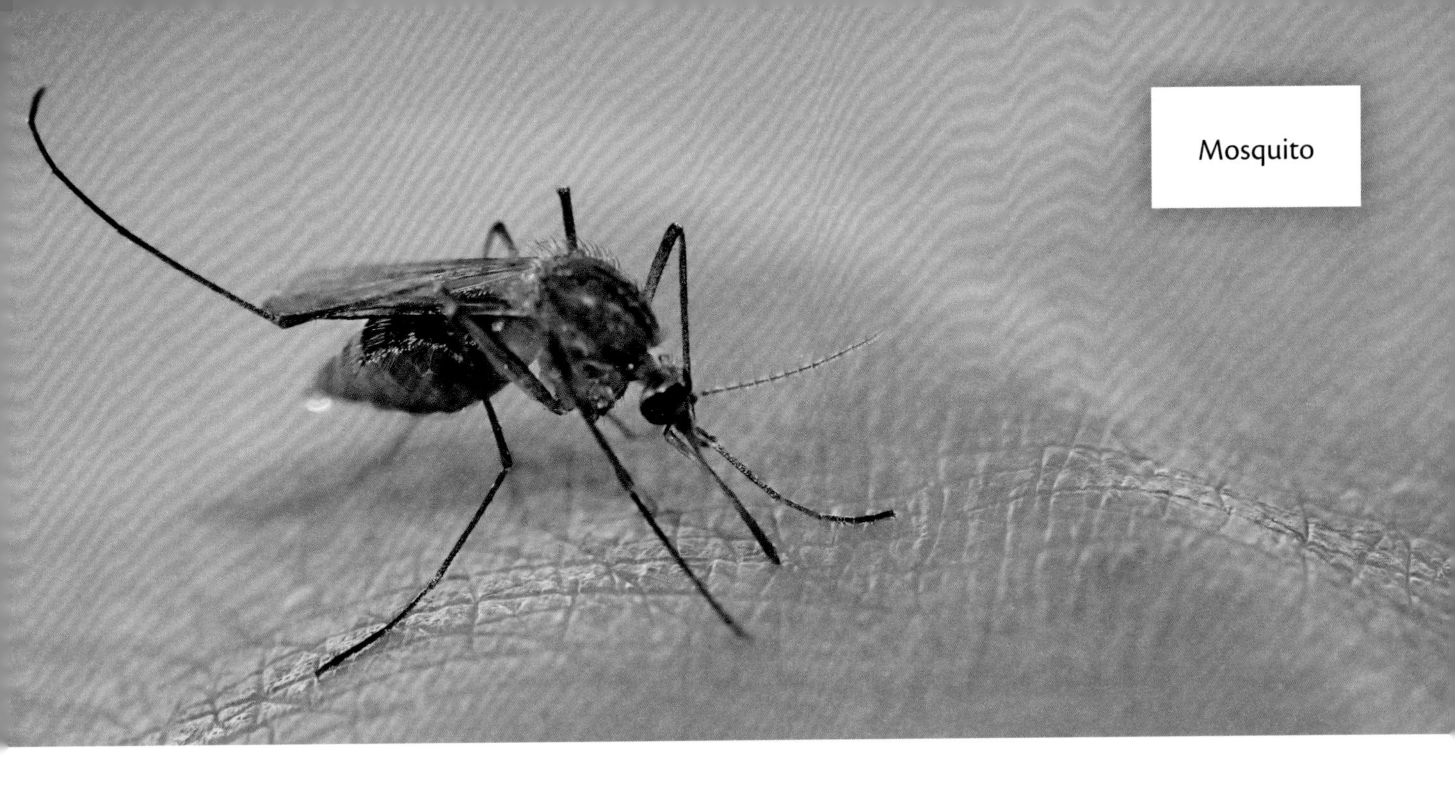
Mosquito

Creatures that bite to drink your blood:

- Mosquitoes
- Some Flies
- Lice
- Fleas
- Bed Bugs
- Ticks

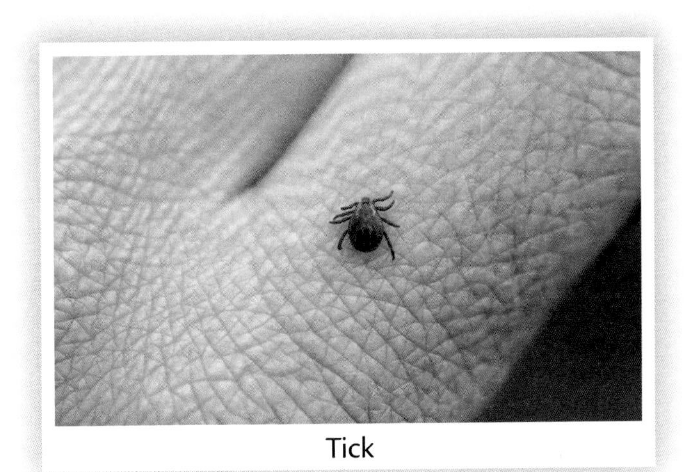
Tick

Creatures that eat your stuff:

- Beetles and moths that eat fur, wool, and silk
- Beetles and moths that eat stored grain
- Cockroaches and silverfish that eat anything
- Carpenter bees, carpenter ants, and termites that eat wood
- Many kinds of bugs that eat crops and gardens

Termites

Creatures that sometimes spread disease:

- Mosquitoes, fleas, ticks, lice, flies, and South American bed bugs can spread diseases when they bite.
- Flies carry germs on their spongy mouthparts and feet.

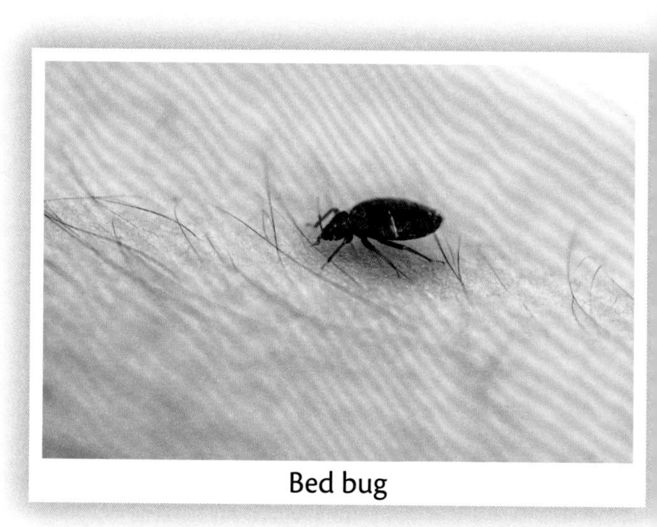
Bed bug

Ew. Now I feel itchy after talking about all these yucky pests. Let's pray and then do a fun activity that's not about pests!

Prayer

Thank You, God, that we hardly ever see or get bitten by these pests. We pray for You to protect people from sickness and ruined food caused by them. Thank You for making interesting creatures that can breathe in the water. Amen.

Time to do Activity 60 in the Activity Book!

A corn earworm feeds on a corncob.

Yellow moray eel

UNIT 6

Animals with Amazing Skin

Now we're going to learn about fish, amphibians, and reptiles! God made each of these creatures with everything they need to live. Our memory verse shows that He cares for our lives too!

Our hymn reminds us that we can see God's love in His creation!

Memory Verse

"In [God's] hand is the life of every living thing
and the breath of all mankind."
Job 12:10 (ESV)

Hymn Singing

A Little Child May Know

A little child may know
Our Father's name of love;
'Tis written o'er the earth below
And on the sky above.

Around me when I look,
His handiwork I see;
This world is like a picture book
To teach His love to me.

The birds that sweetly sing,
The moon that shines by night,
With every tiny living thing
Rejoicing in the light,

And every star above,
Set in the deep blue sky,
Assure me that our God is love
And tell me He is nigh.

This hymn may be sung to the tune of "The Farmer in the Dell," which can be found on the internet.

CHAPTER 21

Fish and Other Underwater Creatures

Then God said, "Let the waters abound with an abundance of living creatures, . . ." So God created great sea creatures and every living thing that moves, with which the waters abounded, according to their kind. Genesis 1:20-21

Fish with Bones

Most of the fish in the world have skeletons made of bones. Most of these bony fish:

1. Have hard, round scales to protect their skin
2. Have gills
3. Are cold-blooded
4. Have babies that hatch from eggs
5. Are given fins for swimming
6. Use swim bladders inside them to raise and lower themselves in the water

Powder blue surgeonfish

To help fish swim quickly, God made their skin smooth by covering their scales with slime.

Scales are small, hard plates that grow out of a fish's skin. They protect the fish and give it color.

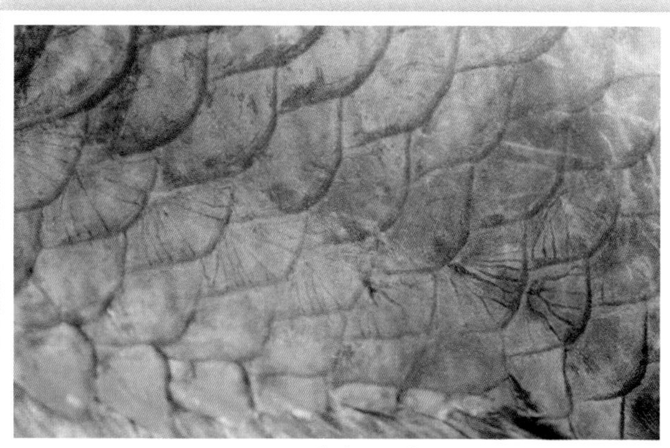

Parrotfish scales

God made a way for fish to feel with their skin even though it's covered with **scales** and slime. They have a line of tiny hairs on their sides. These special hairs help the fish know when something is making the water move near them. Without this line of hairs, a fish would not be able to catch food or

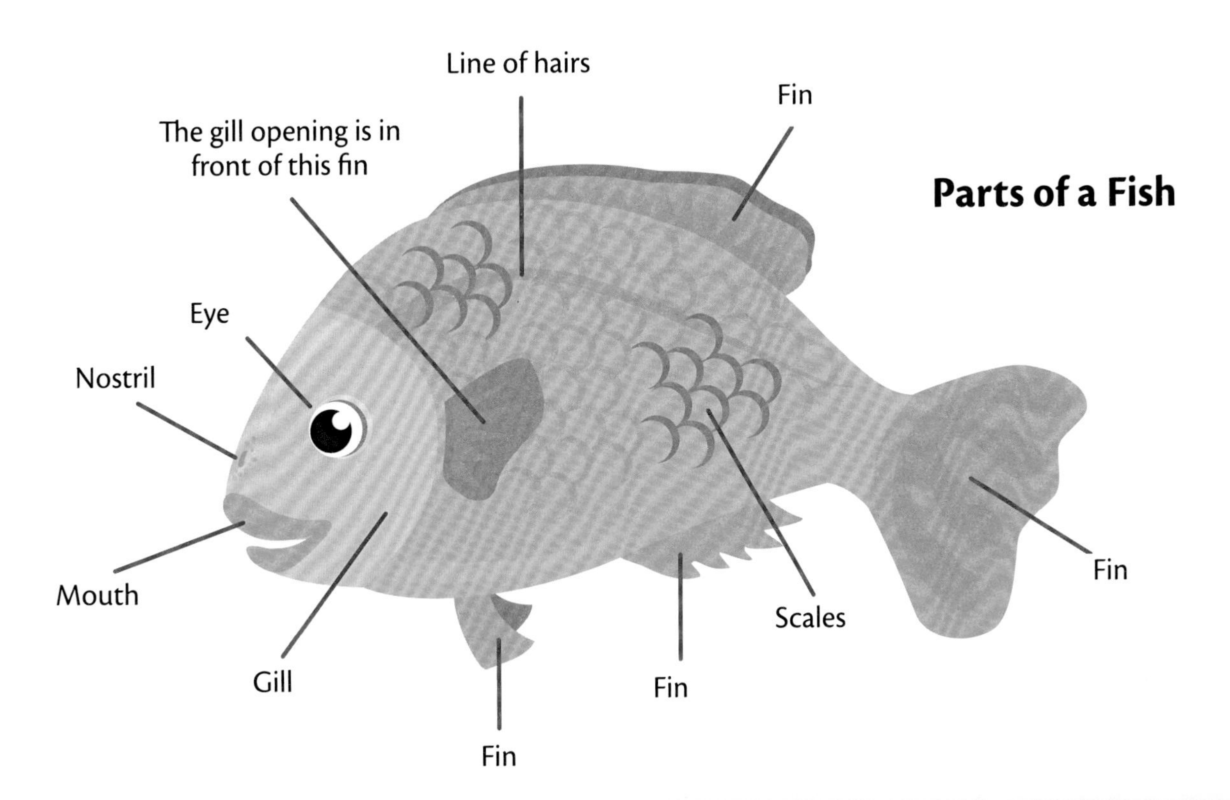

know if an enemy was nearby. When fish swim together in big groups, their lines of hairs help them all move and change direction together.

Frillfin goby

Most water creatures use gills to get oxygen from the water. A fish will open its mouth to bring in water. Then it pushes the water over its gills to bring oxygen into its blood. The gills also take carbon dioxide out of its blood and put it into the water. As the fish closes its mouth, that water goes out through openings in the sides of its head.

Tide pools are puddles that have been left behind in the rocks when the tide is low. The **frillfin goby** is a smart little fish that lives in tide pools. The goby cannot swim away if a hungry octopus or bird is after it. What does it do? It jumps out of its pool! But how does it know it won't land on the rocks? When the tide is high, it swims above all the rocks. It memorizes where the pools will be at low tide. Then it

Yellow tang

Striped bass

knows where it can jump to escape from an enemy! A frillfin goby can memorize where all the tide pools are after seeing them only once. And it can remember this forty days later![15]

Time to do Activity 61 in the Activity Book!

Fish with Cartilage

Some fish have skeletons inside them that are not made of bones. Instead, their skeletons are made of **cartilage**.

Sharks, **skates**, and **rays** have skeletons made of cartilage. They are all cold-blooded. Skates and rays are flat with skinny tails. They look like they are flying as they swim. Rays have dangerous spines on their tails. Their babies are born live instead of hatching from eggs as skates do.

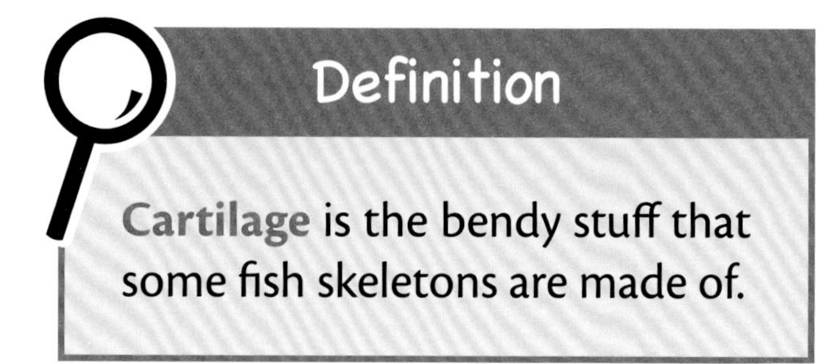

Definition

Cartilage is the bendy stuff that some fish skeletons are made of.

Belly of cownose ray

Shark

Sharks are shaped like most fish. They do not have swim bladders, so they must keep swimming to stay a certain depth. A few kinds of sharks can only breathe by swimming with their mouths open. This brings in water for their gills. A few other kinds of sharks don't swim much. They can only breathe by pushing water through their gills with their mouth muscles. Most sharks can breathe both ways when they need to.

A baby shark will hatch inside this shark's purse (egg case).

Some sharks lay eggs that they attach to something in a safe place. Other shark eggs hatch inside their mother and stay attached to her until they are born. Some others hatch inside their mother and eat other eggs and babies inside of her until they are born.

God gave sharks an amazing way to find food. They find an animal to eat by feeling the electricity its muscles put out when it moves.

Time to do Activity 62 in the Activity Book!

Sea Creatures without Skeletons

Many animals that live in water do not have skeletons. Instead, God gave each kind of creature a way to keep its shape. Here are some of them.

Sponges hold their shape with tough, sturdy webs all through their bodies. Before people made plastic sponges to clean with, fishermen would gather real sponges from the bottom of the ocean. After the sponges died and only the webs were left, people used them to soak up liquid into the many holes. Sponges cannot move. To eat, they push water through tubes in their bodies and trap the tiny creatures that come with it.

Coral

Jellyfish

Anemone

Jellyfish, **corals**, and **anemones** do not have gills. Their skin is so thin that oxygen and carbon dioxide can pass right through it. They all have tentacles to catch and sting their food. They are able to keep their shape because water holds them up. Corals build little stony cups to live in. Large colonies of coral creatures are built on top of the cups of dead ones.

Mollusks are animals that make their own seashells to live in. Their shells are as hard as stone and often very beautiful. Some creatures make their shell like an upside-down bowl that they live under. **Snails** make their shells in a spiral with a little round door they can close over the opening. **Clams** and **oysters** make two matching shells that they live between. They can squeeze their home together with a very strong muscle. These shells protect them from their enemies.

Fish fill the oceans and the seas. They are a beautiful part of God's creation. We can't control the fish or tell them what to do, but God can!

A living Scotch bonnet mollusk with its shell on its back

Now the LORD had prepared a great fish to swallow Jonah. And Jonah was in the belly of the fish three days and three nights. (Jonah 1:17)

Jesus said, ". . . go to the sea, cast in a hook, and take the fish that comes up first. And when you have opened its mouth, you will find a piece of money." (Matthew 17:26-27)

And [Jesus] said to them, "Cast the net on the right side of the boat, and you will find some." So they cast, and now they were not able to draw it in because of the multitude of fish. (John 21:6)

Prayer

Thank You, God, for the wise little goby and for the good job gills do. We praise You for beautiful sea shells that give a safe covering for the creatures inside. Amen.

Time to do Activity 63 in the Activity Book!

CHAPTER 22

Amphibians

Amphibians are the wonderful creatures we call frogs, toads, and salamanders. Another one of the 10 big problems that God sent to the Egyptians was a plague of amphibians.

[God said to the king of Egypt,] "But if you refuse to let [My people] go, behold, I will smite all your territory with frogs. So the river shall bring forth frogs abundantly, which shall go up and come into your house, into your bedroom, on your bed, into the houses of your servants, on your people, into your ovens, and into your kneading bowls." (Exodus 8:2-3)

Amphibians:

1. Are cold-blooded
2. Have a line of hairs, like fish do, until they are adults
3. Hibernate by burying themselves in underwater mud or in the ground

Well, the king of Egypt didn't let God's people go and God sent the frogs. I think frogs are cute. The Egyptians probably did not think frogs were cute. They couldn't do anything or go anywhere without a slippery frog being there too!

Frogs with eggs

4. Have four ways of breathing
5. Shed their skin and eat it
6. Don't have ear openings but hear through circles of tight skin
7. Start their lives as water creatures
8. Can live on land when they grow up

Frogs can get oxygen through the skin of their mouths.

Amphibian Metamorphosis

God made the bodies of amphibians to go through big changes as they grow up. We call these big changes metamorphosis, as we do with insects. Here are the steps of frog metamorphosis:

1

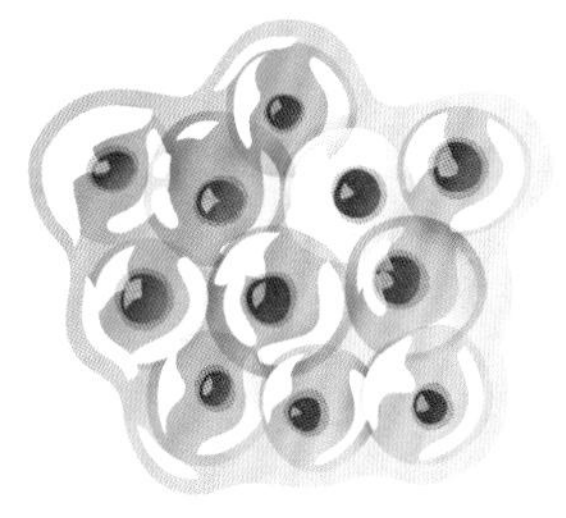

Egg

Most mother amphibians lay their eggs in water.

2

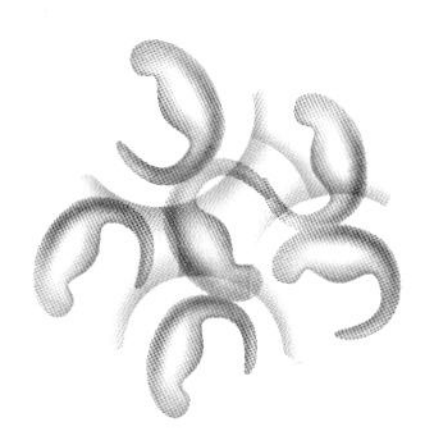

Embryo

The babies grow body parts inside the eggs.

3

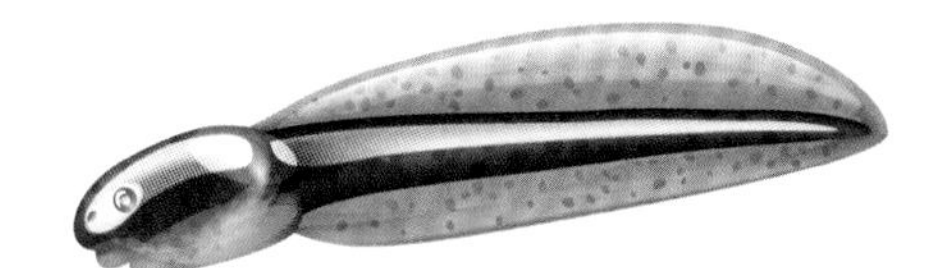

Tadpole

Baby frogs hatch out of the eggs. They look like heads with tails attached. They have gills for breathing.

4

Tadpole with two legs

Tadpole legs grow under their skin. The back legs are the first to pop out.

5

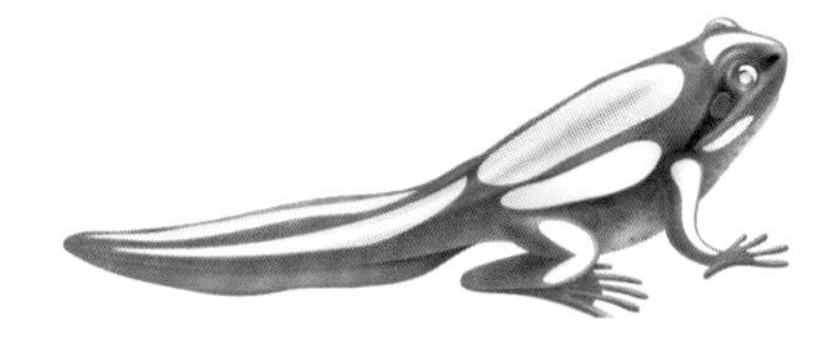

Tadpole with four legs

The front legs pop out next. The tail begins shrinking.

6

Adult frog

The tail shrinks until it is gone. Adult frogs have lungs. Their gills disappear.

Adult frogs can get oxygen from air through their lungs and through the skin inside their mouths if they hold them open. But frogs get most of their oxygen through the skin on their bodies. Their skin can get it from air or from water. The blood inside their bodies is very close to their skin. Their skin is thin and stays moist so that oxygen can easily go through to the blood. When frogs hibernate and bury themselves, they only breathe through their skin. When oxygen comes into an amphibian's blood, carbon dioxide goes out through its gills, lungs, mouth, and skin.

Time to do Activity 64 in the Activity Book!

Frogs and Toads

Frogs and toads have long back legs for jumping. If a frog were as big as you, it would be able to jump 80 feet (25 m)! When male frogs or toads want to find females, they will make loud noises to help the females find them. Some croak, some peep, and some trill. Bullfrogs sound like a cow! The sounds are made with their voices, but the sounds become louder because the frogs blow up a big bubble on their throats.

Frogs spend most of their time in water. God made their back feet webbed to help them swim quickly. Frogs have smooth skin that keeps itself moist. When a frog sees a tasty insect, it jumps and grabs it with its sticky tongue. To swallow, the frog has to squeeze its eyes shut and push them into its head. The backs of the eyeballs help push food down its throat.

Male frog calling a female

Frogs have webbed feet.

Toads spend most of their lives on land. They only go in water when it's time to lay eggs. God did not give toads webbed feet. He did give them a way to protect themselves. Behind each eye is a place where poison squirts out if an animal bites the toad. The poison makes the animal feel sick and dizzy. It tastes so bad that the animal will remember to never bite another toad! Toads can catch insects without jumping. Instead, their tongues do all the work. Toad tongues are attached to the front of their mouths. Their tongues are very long, very quick, and very sticky. Toad tongues shoot out to catch their food. Like frogs, their eyeballs help them swallow it. The skin of a toad is dry and bumpy.

Toad catching an insect with its sticky tongue

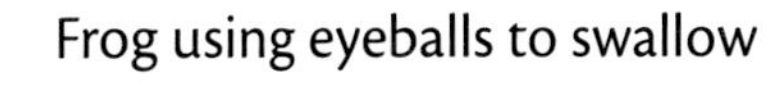
Frog using eyeballs to swallow

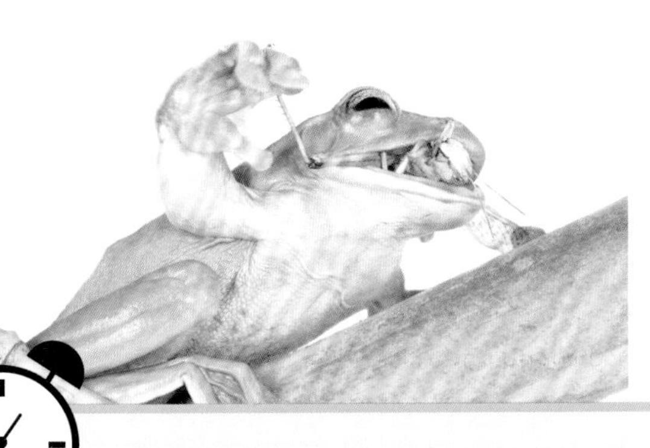

Salamanders

Salamanders are amphibians that do not lose their tails when they become adults. Some adult salamanders live in water and use their tails for swimming. Other adult salamanders live in trees and use their tails to wrap around branches like monkeys do. Most salamanders live on land.

Colorful salamanders make poison on their skin. The tail has extra poison on it. If an enemy comes near, the poisonous salamander wiggles its tail. An enemy looking for food will usually bite something that looks alive and wiggly.

Poisonous salamander

Nonpoisonous salamander

When the enemy bites the tail, the salamander makes it come off in the animal's mouth! Then the salamander drops safely to the ground and scurries away. The animal spits out the tail and never tries another taste of salamander. What does the salamander do without a tail? It soon grows a new one! Salamanders can also grow new legs or new eye parts if they lose them.

[Jesus] said to the man, "Stretch out your hand." And he did so, and his hand was restored as whole as the other. (Luke 6:10)

And one of them struck the servant of the high priest and cut off his right ear. But Jesus . . . touched his ear and healed him. (Luke 22:50-51)

Wow! God can make new parts for salamanders! Jesus gave a new hand to one man and a new ear to another. Of course the Creator can heal!

During metamorphosis, a salamander grows eyelids, teeth, and a tongue. Land salamanders lose their gills. Unlike frogs and toads, the legs that salamanders get during metamorphosis are all about the same size. Most salamanders do not have lungs as adults. They breathe through their moist skin.

Salamanders do not use noises to find each other. They use smells instead. Salamanders go into the water to lay their eggs. Some mother salamanders guard their eggs until they hatch.

Prayer

We praise You, God, for Your amazing amphibians! They are kind of cute and kind of funny. Their bodies can do such interesting things. They can even grow back missing parts. Amphibians are special in Your creation. Amen.

Time to do Activity 66 in the Activity Book!

Horned toad warming itself

CHAPTER 23

Reptiles

Snakes, lizards, turtles, tortoises, alligators, and crocodiles are all reptiles. Just like the other animals we have learned about in this book, their bodies do not make heat. They stay the same temperature as the air or water around them. As soon as the sun comes up in the morning, reptiles move to a sunny spot to lie in. They are not content to stay in a cold, dark place. They want to be warm!

Then Jesus spoke to them again, saying, "I am the light of the world. He who follows Me shall not walk in darkness, but have the light of life." (John 8:12)

How Are Reptiles Special?

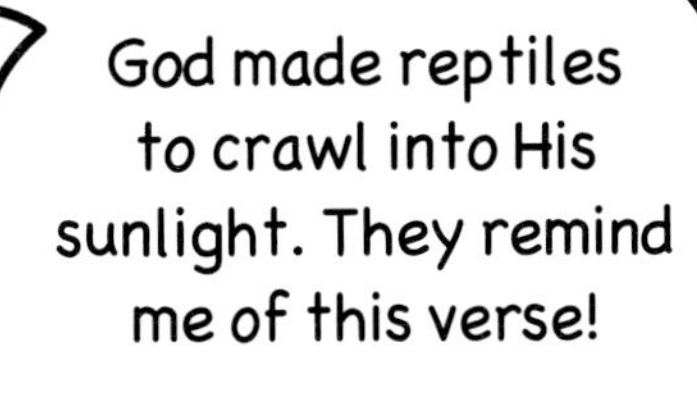

Reptiles:

1. Are covered with tough scales that keep them from drying out
2. Shed their skin as they grow
3. Breathe only with lungs
4. Lay eggs that have tough, bendable shells

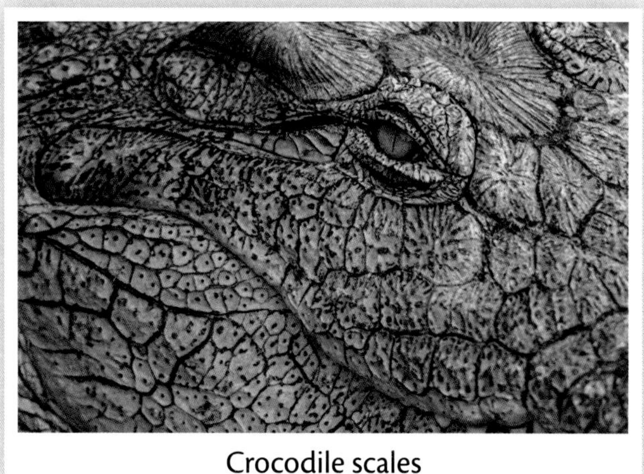

Crocodile scales

A snake sheds its skin.

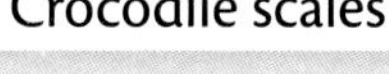

Time to do Activity 67 in the Activity Book!

Snakes and Lizards

Snakes are a reptile that many people are afraid of. Some have poisonous bites, so it's good to be careful when you are outside where they live. Most snakes are not dangerous. All snakes are helpful because they eat mice and other pests.

The most unusual thing about snakes is that they don't have legs! God gave them scales on their tummies that are able to grab the ground. Snakes move by twisting or stretching themselves forward while their tummy scales are grabbing.

Snakes cannot see or hear very well. But God has given them an amazing sense of smell. They can smell better than any other animal.

Sea turtle hatching

This helps them find their food. Snakes don't smell through their nostrils. They smell with their tongues by flicking them out of their mouths.

Some snakes have two holes, called pits, between their eyes and their nostrils. These pits help them feel the temperature of things around them, especially warm-blooded animals to eat. Some snakes bite to kill the animals they hunt.

Snake eating a mouse

Rattlesnake tongue and pits

Others wrap their bodies around the animal and squeeze it until it dies. **Garter snakes** are the only snakes that kill their food by both biting and squeezing.

The poison from poisonous snakes goes into the animal through two long teeth called **fangs**. The poison either kills the animal or makes it unable to move. Snakes cannot chew their food. They swallow animals whole. Snakes can swallow animals that are fatter than they are. How do they do this? The bones in a snake's mouth can come apart to make its jaws open very wide!

Fangs are a snake's two long teeth that put poison into the animal they bite.

In the Bible, God tells us that we should rule over animals and care for His

earth. One way we can do this is by getting rid of poisonous snakes living near our home. We should allow the safe snakes to stay nearby so they can eat pests.

Poisonous snakes can be scary. But remember that when you are afraid, you should put your trust in God. He is the One who protects you.

God made all things good. But when Adam sinned, some animals became dangerous. This is because evil came into the world. Evil things can be very scary. But when we trust in God, we don't need to be afraid of evil. Jesus rules over all things. He crushed the devil when He died on the cross, and He will destroy all evil one day.

Lizards are able to hear better than snakes. They have ear openings just below the skin. And they can walk because they have four legs. The **horned toad** is not really a toad. It's a gentle lizard that can be a good pet. But if your dog attacked

Chameleons have long tongues for catching insects.

this lizard, your horned toad could squirt blood out of its eyes! That would surprise your dog and give the horned toad time to run away.

Chameleons have different layers of skin. The outside layer is clear. The layers under it have tiny crystals that reflect light in different ways. When a chameleon is calm and resting, it is green. That means that the tiny crystals are close together. When a chameleon gets scared or excited, the crystals move apart, and it turns yellow or other bright colors. When chameleons are cold, their skin can get darker to warm them up. Dark colors soak in more of the sun's heat than light colors do.

Some chameleons can change colors differently to protect themselves from different enemies. Birds can see color very well. If the chameleon's enemy is a bird, it carefully matches its skin to whatever it is sitting on so the bird won't see it. Snakes cannot see color very well. A chameleon who sees a snake coming does not change its color as much. But the snake still has a hard time seeing it. How does a chameleon know the way birds and snakes see colors? They know because God wisely gave them this gift.

Time to do Activity 68 in the Activity Book!

Other Reptiles

Turtles and **tortoises** have bodies that are attached to the inside of their hard shells. It is impossible for the shells to come off. Their shells are made up of their backbones, ribs, and other bones. The outside of their shell is covered with the

Sea turtle

Turtle bones

Tortoise

same thing that your fingernails are made of. Shells grow as the turtle or tortoise does. They are not shed like their skin. The turtle usually keeps its head, tail, and legs outside its shell, but it can pull them inside if an enemy comes near.

Some turtles and tortoises can close their shell openings with hinged doors. They hold the doors shut with very strong muscles.

You might think turtles and tortoises look the same, but they are very different. Here are some things that make them different:

Tortoises live on land

and almost never go into water. Turtles live in water most of the time. Tortoise shells are rounded like half of a ball. Turtle shells are flatter, like only the top of a ball. Tortoises have thick legs like elephants so they can carry their heavy shells on land. Turtles either have legs that are flippers or toes that are webbed for swimming.

Alligator

Alligators and **crocodiles** live in and near water. They swim quickly by tucking in their legs and wiggling their tails back and forth. They eat fish, turtles, and sometimes land animals that come to the water to drink. They lie underwater with only their nose, eyes, and ear bumps sticking up in the air. They are hoping an animal will think they are logs and will get close enough to grab! Alligators and crocodiles can stay underwater without breathing for an hour. They spend the night floating in the water and come out on land in the morning to warm up in the sun.

Crocodile

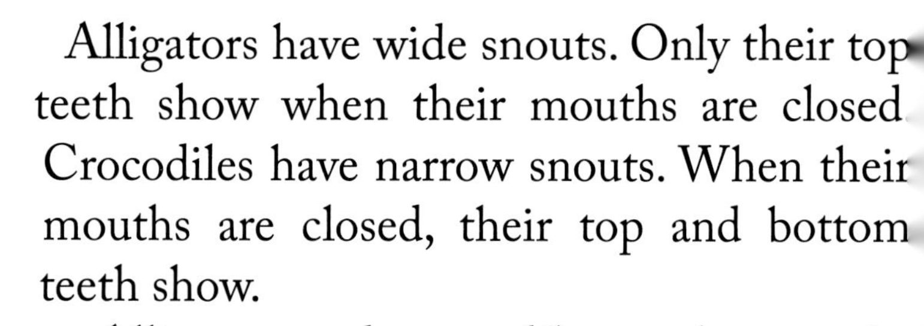

Alligators have wide snouts. Only their top teeth show when their mouths are closed. Crocodiles have narrow snouts. When their mouths are closed, their top and bottom teeth show.

Alligator and crocodile mothers make nests of mud and rotting plants. They bury their eggs in the nest. Rotting plants put out heat and keep the eggs warm. The mothers don't stay near their nests, but they check back now and then to listen for the peeps the babies will make when they hatch. When the mother hears the babies, she will dig them up and carry them to the water.

For with You is the fountain of life;
In Your light we see light. (Psalm 36:9)

Prayer

Reptiles are amazing, God! We like to see them rejoicing in the light. You made their bodies with everything they need to live. Thank You for making them so interesting. Please protect people from the reptiles that are dangerous. And help us to love Your light. Amen.

Time to do Activity 69 in the Activity Book!

CHAPTER 24

Strange Reptiles

Dinosaurs

Dinosaurs were strange and amazing reptiles! Because **dinosaurs** walked on land, we know that God made them on the sixth day of creation with the other land animals. We don't know of any dinosaurs that are still alive today, so we say they are **extinct**.

The Bible says that God brought two of every air-breathing creature to Noah before the flood. That means that dinosaurs were on the ark. Some dinosaurs were huge! How could they fit on the ark? Reptile babies are very small when they hatch from eggs. But they don't stop growing until they die. The dinosaurs on the ark were probably young and small. Their parents may have been very large dinosaurs that

Dinosaurs babies were small.

Definition

Dinosaurs are extinct reptiles that once walked on land.

An animal is **extinct** if it once lived but we think that none of them are alive now.

Fossilized footprint

died in the flood and became fossils.

Many fossils have been found all over the world. Fossils help people know what dinosaurs and other creatures looked like. Dinosaur fossils tell us what shape and size the dinosaurs were. Fossil footprints tell us their weight and how they walked.

Time to do Activity 70 in the Activity Book!

We can learn a little about dinosaurs from pictures and writings from long ago. After the flood, people around the world still saw dinosaurs as they were becoming extinct. In countries all over the earth, there are stories and pictures of people fighting dinosaurs.

Fresco, *St. George and the Dragon*, Monastery Church (Bucovina, Romania) 1537

"Look now at the behemoth,
which I made along with you;
He eats grass like an ox.
See now, his strength is in his hips,
And his power is in his stomach muscles.
He moves his tail like a cedar [tree];...
Indeed the river may rage,
Yet he is not disturbed."
(Job 40:15-17, 23)

There are even a few places in the Bible that seem to be talking about dinosaurs!

Argentinosauruses

The **behemoth** may have been one of the long-necked dinosaurs. Fossils show that this kind of dinosaur also had a very long tail like a cedar tree. Its legs were like tree trunks.

Tyrannosaurus rex

The scariest dinosaurs had large, sharp teeth. Fossils show that these dinosaurs could bend and twist their heads very far in all directions. They walked on two strong back legs. Their arms were small compared to the rest of their body.

Torosaurus

Other fossils show us dinosaurs with big horns and shields on their heads.

Young and old stegosauruses

Some dinosaurs looked a little like tortoises. They had short legs and rounded backs. Some of them even had bony bumps or flat plates all over their backs. Others had spikes or plates sticking up along the top of their backs. Some of these turtle-like dinosaurs had spikes or clubs on their tails.

Time to do Activity 71 in the Activity Book!

Flying Reptiles

When many dinosaurs lived on the earth, there were also strange reptiles in the air. Because they could fly, these reptiles are not actual dinosaurs. God made flying reptiles on the fifth day of creation when He made birds.

In the Bible, Isaiah might have been talking about flying reptiles when he calls Egypt a land of trouble. He says that lions, snakes, and fiery flying serpents came out of Egypt (Isa. 30:6).

Fossils show that the wings of flying reptiles were not like bird wings. They were made of skin like bat wings are. The front of each wing was attached along the arm. At the wrist, three claws came out. These claws were used as fingers. From the wrist to the wingtip, the wing was attached to the reptile's very long fourth finger. The back edge of the wing stretched from the reptile's ankle to the tip of that long finger.

There were two groups of flying reptiles. One kind was small. This kind had a lot of teeth and a long tail.

The other kind was large. If these flying reptiles stood on their feet like birds, some of them might have been as tall as giraffes. Their wings may have been 36 feet (11m) long from tip to tip. They had short tails, long necks, and large heads. Some of them had crests on their heads.

Plesiosaur

Ocean Reptiles

Fossils also show us big ocean creatures that seem to be reptiles. These would not be called dinosaurs because they swam instead of walked. They had flippers instead of legs. Ocean reptiles would have been created on the fifth day of creation when God made the great sea creatures.

Some ocean reptiles had long necks. Some had short necks and looked like huge crocodiles. Others looked like giant lizards. People have also found reptile fossils of sea creatures that looked like 50-foot (15 m) porpoises.

Three places in the Bible talk of huge, fierce sea creatures called **Leviathans**. Leviathans show us how powerful God is since He can create such big, scary animals.

This great and wide sea, in which are innumerable teeming things,
Living things both small and great.
There the ships sail about;
There is that Leviathan which You have made to play there.
(Psalm 104:25-26)

Prayer

Lord, we are amazed at the huge and strange creatures that once lived on Earth. You are so powerful! Thank You for the reptiles that swam. Thank You for the ones that flew. And thank You for the dinosaurs too. Amen.

Time to do Activity 72 in the Activity Book!

Giant ichthyosaurs probably looked like porpoises.

Dilophosaurus

God cares
for sparrows
and for you!

UNIT 7

Feathered Friends

It's going to be really fun to learn about birds! God gave special gifts to birds because He wanted them to fly!

Our memory verse tells us how we can be strong in the Lord like an eagle.

Our hymn sings about God's care of sparrows. Since He cares for them, we know He cares for us!

Memory Verse

But those who wait on the LORD
Shall renew their strength;
They shall mount up with wings like eagles.
(Isaiah 40:31)

Hymn Singing

God Sees the Little Sparrow Fall

God sees the little sparrow fall,
It meets His tender view;
If God so loves the little birds,
I know He loves me too.

Chorus:
He loves me too, He loves me too,
I know He loves me too;
Because He loves the little things,
I know He loves me too.

He paints the lily of the field,
Perfumes each lily bell;
If He so loves the little flow'rs,
I know He loves me well.

God made the little birds and flow'rs,
And all things large and small;
He'll not forget His little ones,
I know He loves them all.

You can listen to this hymn by searching for "God Sees the Little Sparrow Fall, children's hymn" on the internet.

CHAPTER 25
God's Gift of Flight

Birds are amazing! We can enjoy birds with each of our five senses. We have fun watching them busily search for food. We can enjoy hearing the music of songbirds early in the morning. We feel soft goose feathers in fluffy pillows and jackets. And now and then, we might get to enjoy a roasted chicken on the table. That's a treat for our nose and mouth! But when we think about birds, we usually think of their amazing flight.

Feathers and Bones Are Built for Flight

Birds were built by God with special parts that make them able to fly. Do you remember when we learned about flight in Chapter 9? We learned that bird wings have a special shape. That shape is important as a bird flies. It makes the air move faster over the top of the wing and slower underneath. That's what makes the wing lift up and helps the bird fly!

Part of what gives a bird's wing its shape is its feathers. The first feathers a baby bird gets are fluffy. As it grows, it gets new feathers on top. But it keeps some fluffy feathers next to its skin for its whole life. These feathers help keep birds warm. They are called **down feathers**.

Adult bird feathers are waterproof! Mother and father birds spread their wings over their babies in the nest. This keeps the babies dry until they grow waterproof feathers of their own. Birds help us understand this verse!

[God] shall cover you with His feathers,
And under His wings you shall take refuge.
(Psalm 91:4)

The feathers on top of a bird's down feathers are flat. The flat feathers make the bird's body smooth so it can move easily through the air. Each flat feather is very thin. It's attached to the bird's body by its hollow quill. On each side of the quill are many tiny branches.

Baltimore oriole taking off

The large, flat feathers on a bird's wings and tail are called **flight feathers**. Most flight feathers are strongly attached to the bird's bones. Birds use the flight feathers on their wings to push air while they fly. The feathers on their tail are used for steering.

Flight feathers have a special job. They must keep air from passing through them as the bird flies. The branches of flight feathers are attached to each other by tiny hooks. If the branches become unhooked, air will pass through the feather, and the bird will have trouble flying. The bird uses its beak to zip the branches together again if they get unhooked. Birds spend a lot of time fixing and cleaning their feathers. This is called **preening**. Many kinds of birds have a place near their tail that makes special oil. Birds use their beaks to spread this oil on their feathers.

When something doesn't weigh very much, we say it's as light as a feather. We say that because feathers are very light. But all of a bird's feathers are still heavier than all of its bones! God made bird bones very light.

If a bird had heavy bones like a person, it would not be able to fly. God made bird bones light by making

Preening parrot

Definition

Preening is when a bird uses its beak to fix, clean, and oil its feathers.

Bridge with support structures like bird bones

them hollow. Even though they are hollow and filled with air, they are still very strong. The inside of the bones are crisscrossed with supports made of bone These supports are like the supports that hold up a bridge. They are made just right to give bird bones the most strength with the lightest weight.

Time to do Activity 73 in the Activity Book!

A Bird's Breathing Is Designed for Flight

All animals need oxygen to live. Birds use more oxygen than any other animal Flying is hard work. When your body works hard to run fast, you need to take a lot of deep breaths to get more oxygen. God gave birds a very special way of breathing that lets them get more oxygen out of the air than any other animal can. They can even fly high up in the sky where there is not much oxygen at all.

Bird lungs bring oxygen in from the air. Then their lungs take carbon dioxide out of their bodies. But birds also have nine air sacs that fill with air when they breathe. The air sacs help move the air through their lungs twice to get the most oxygen out of it.

Sometimes birds seem to sing a long time without taking a breath! That's because they use the air in their air sacs. They also take mini-breaths that are too short for us to hear!

Time to do Activity 74 in the Activity Book!

A Bird's Eating Is Designed for Flight

We've learned about many animals that are cold blooded. Their bodies slow down in the cold. But birds cannot do that. There is no way to fly slowly without falling to the ground! So God made their bodies to stay the same temperature no matter how cold or warm the air is. They are **warm blooded**. Because they are warm blooded, they can fly in winter and summer and be busy day or night. Birds must eat a lot of food to make energy to warm their bodies.

God gave birds the ability to eat quickly. Then they can go somewhere safe while their body uses the food. They can also use the food they eat much sooner than other animals can. This gives them quick energy for flying.

Definition

Warm-blooded animals are able to keep their bodies the same temperature wherever they are.

Blackbird eating a blueberry

Air travels into a bird's lungs (pink) and air sacs (blue).

The Path of Food through a Bird

1. When a bird swallows some seeds, the seeds first go into a sac called a **crop**
2. Bird **stomachs** are small, so the crop saves the food until the stomach is ready. When the seeds go to the stomach, the stomach adds liquids that make the seeds start to soften.
3. Then the seeds go to the **gizzard**. The gizzard grinds the seeds with its rough inside. Some birds swallow little rocks to help the seeds get ground in the gizzard.
4. The ground seeds go into the **intestine**. The intestine is where energy and vitamins are taken out of the food and put into the blood.
5. Liquid and solid waste go out of the bird through an opening under the tail. When a bird starts to fly, it first lets out its waste to get rid of extra weight.

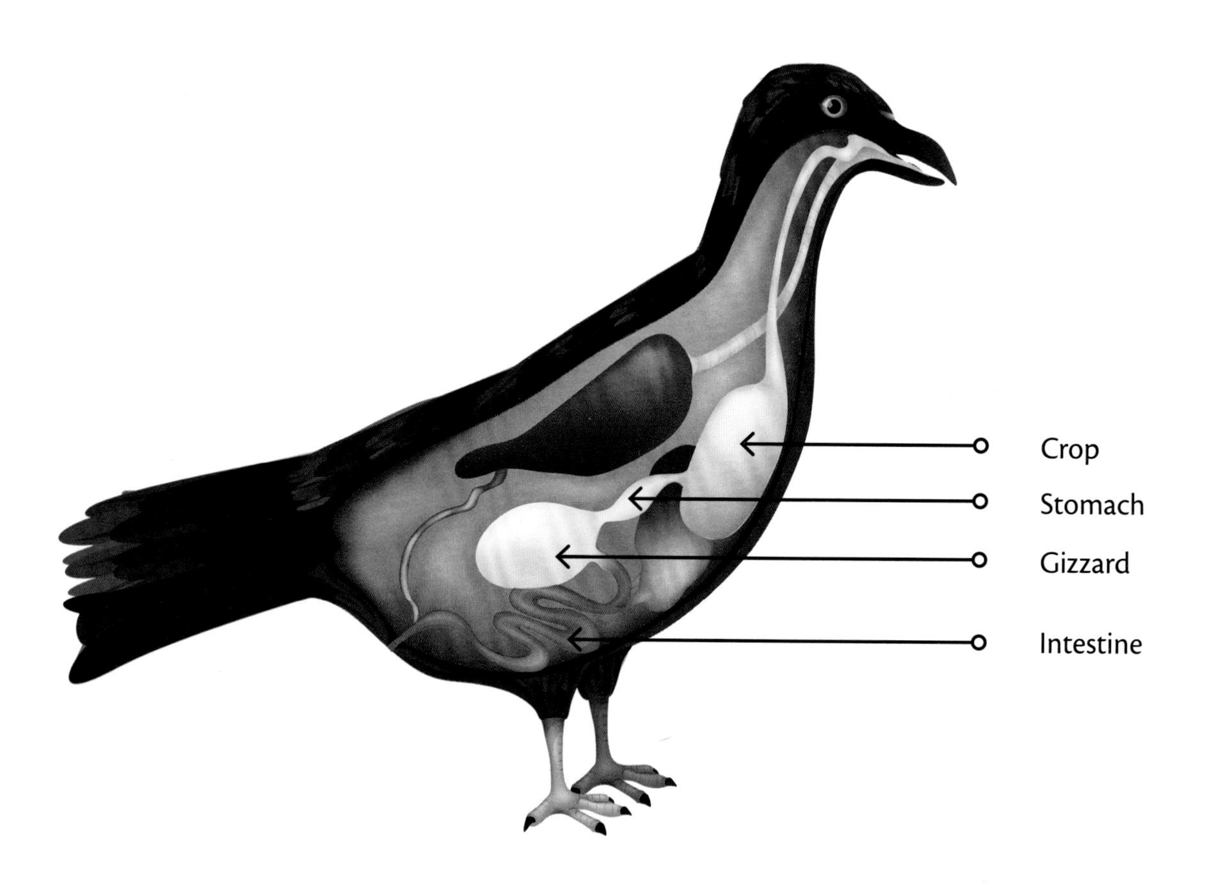

Prayer

Oh Lord, we praise You for the beauty of flying birds. We like to see hawks soaring high. We like to see wrens flitting in the bushes with their little tails up. You made birds able to fly by Your wisdom. Their bodies are built with feathers and bones for flight. You gave them everything they need for flying. Amen.

Time to do Activity 75 in the Activity Book!

Northern harrier hawk

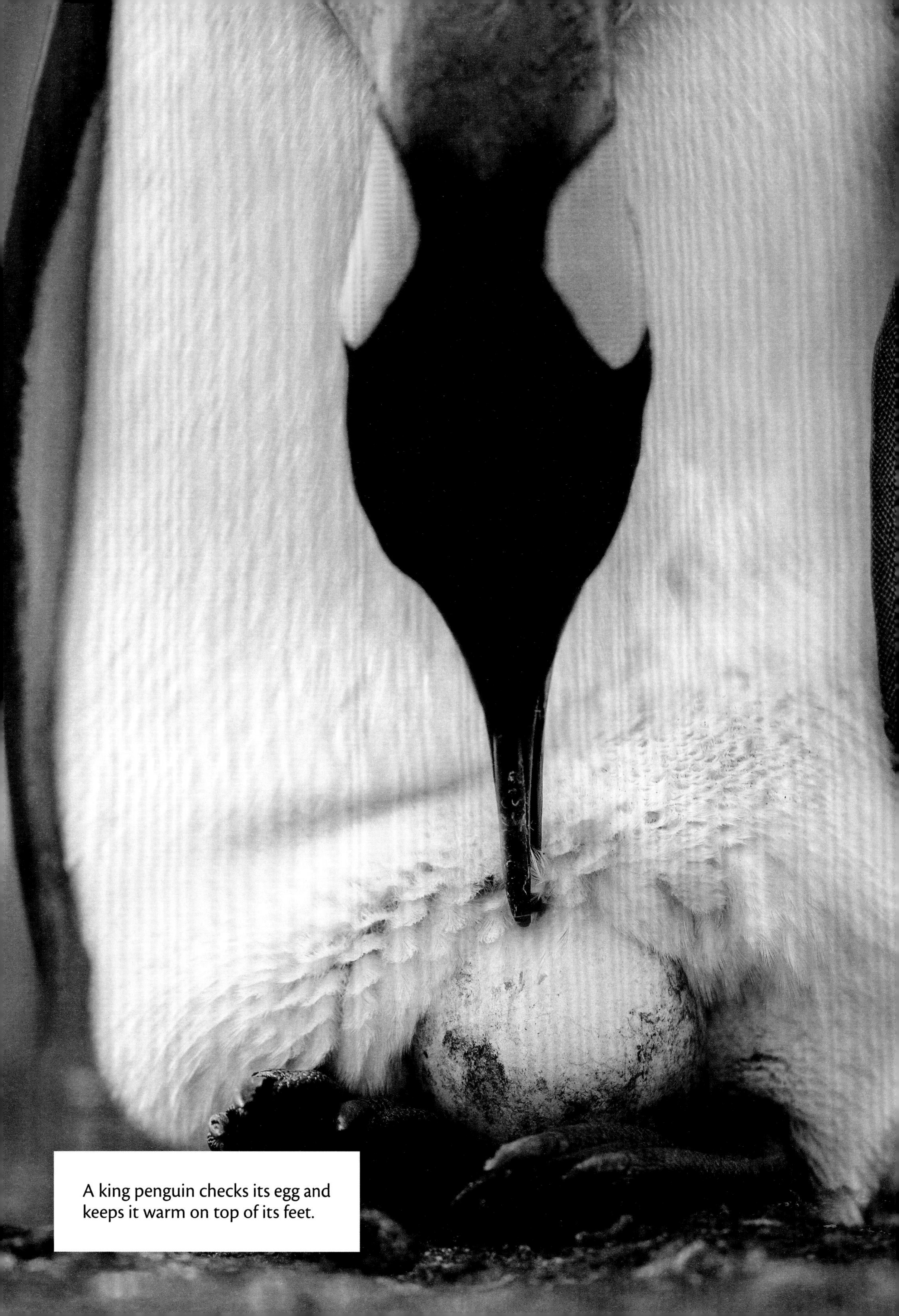

A king penguin checks its egg and keeps it warm on top of its feet.

CHAPTER 26

Beaks, Feet, and Eggs

We have learned about the gifts God gave birds so they can fly. He also gave them other gifts: babies in eggs, special feet, and beaks for eating certain foods. Birds are very different from us! And God made us different from all the animals!

All flesh is not the same flesh, but there is one kind of flesh of men, another flesh of animals, another of fish, and another of birds. (1 Corinthians 15:39)

Beaks Created for Each Kind of Bird

Not all birds eat seeds. Some eat insects. Others eat animals or fish. Some eat plants. Many birds eat a mixture of these things. A bird's beak is its tool for eating. God gave each kind of bird its own special beak.

A toucan's large beak may help the bird stay at a comfortable temperature.

A cardinal's beak is a seed cracker.

An eagle's beak is a meat tearer.

A hummingbird's beak is a flower probe.

A parrot's beak is a nut cracker.

A snakebird's beak is a fish spear.

A pelican's beak is a fish net.

A duck's bill is a water strainer.

A woodpecker's beak is a wood cutter.

Time to do Activity 76 in the Activity Book!

Feet Created for Each Kind of Bird

Most of the birds in the world are perching birds. This means they have special feet that can curl around small branches. Three toes face forward, and one toe faces backward. Their toes automatically curl when the bird's feet are pressed against the branch. That is how perching birds can sleep without falling!

Duck feet

Feet of a bird of prey

Birds that spend most of their time swimming on top of the water have webbed feet. They have skin stretched between their toes to help them paddle faster.

Birds that hunt and fish for their food have sharp claws on each toe. Their feet are very strong so they can hold on to an animal with their claws while flying.

The **ruffed grouse** has fringes on each toe. In winter, its toes give the grouse snowshoes! The fringes spread out when they walk on snow to keep them from falling through.

Ostriches don't fly. They do walk a lot, so God has given each foot special toes to walk on. Each foot has only two toes. One toe is

Ruffed grouse in snow

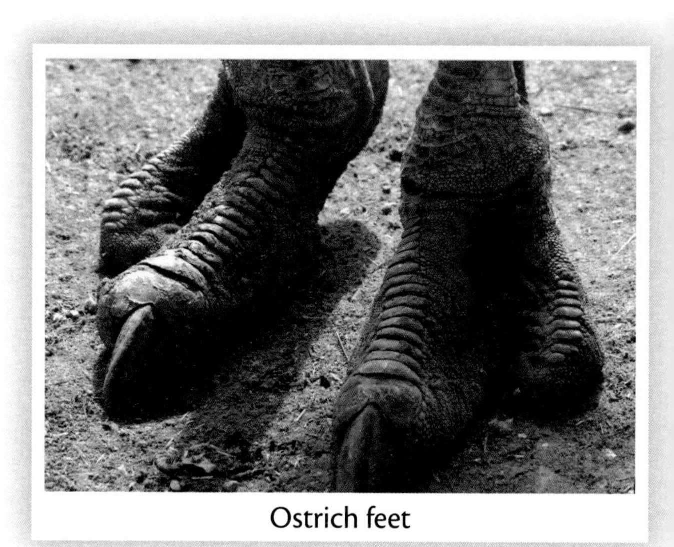
Ostrich feet

much larger than the other. The bigger toe has a wide claw that is like a hoof. Male ostriches can be 9 feet (2.75 m) tall and weigh 330 pounds (150 kg). Ostriches can kick hard enough to kill an adult lion!

Time to do Activity 77 in the Activity Book!

Baby Birds Created in Eggs

Baby birds grow inside of eggs. Why did God make them grow that way? Why don't birds have babies like dogs do? Maybe one reason is that a mother bird couldn't fly very well if she had baby birds growing inside her. She would be too heavy! Also, a mother bird would have to use a lot of her body's energy to grow babies inside. By laying eggs, the babies can grow outside her. She only has

Father robin taking a turn on the nest

to use some of her energy to sit on the eggs to keep them warm. The father bird helps out by sitting on the eggs sometimes and by bringing food to the mother. They share the energy it takes to have babies!

A baby bird is called a **chick**. Not all eggs have chicks growing inside them. God made some birds to lay extra eggs so we would have something to eat. You've probably eaten chicken eggs before. Have you ever eaten an ostrich egg? These eggs are twenty times the size of a chicken egg. One ostrich egg would make enough scrambled eggs to feed your entire family!

Prayer

We praise You, God, for all the different bird feet and bird beaks You made. It's fun to learn what birds eat and do. Then we can see how their special feet and beaks help them! Thank You for making a way for birds to have babies in eggs. Thank You that we can eat eggs and use them to make cookies. Amen.

Time to do Activity 78 in the Activity Book!

Robin chick, soon after hatching

CHAPTER 27

Bird Migration

Migration

Do you remember when we learned about the long trips that monarch butterflies take? Many kinds of birds also take long trips. These trips are called **migration**.

Why do birds migrate? At the end of summer, the chicks are grown up. Then fall weather comes to the nesting area. The insects and plants begin to die in the colder weather. Birds can't find as much food, and they are not content. When a cold storm comes, they decide it's time to go to their warmer winter home. They take a journey, flying to the same place where they or their parents had spent the last winter.

Definition

Bird migration is when a bird moves from its summertime nesting area to a new winter home and back.

When spring comes again, it's time for birds to have chicks. The far-away nesting area has warmed up. A lot of insects are hatching there. Plants are growing again. There will be enough food and enough room to raise a family. So the birds leave their winter home to travel back to where they spent the last spring and summer.

"Even the stork in the heavens
Knows her appointed times;
And the turtledove, the swift, and the swallow
Observe the time of their coming.
But My people do not know the judgment of the LORD."
(Jeremiah 8:7)

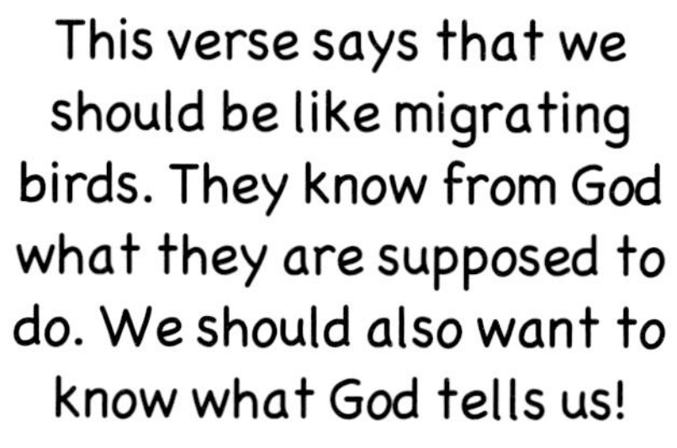

Black swifts

Some birds do not migrate. They have enough of the food they need at the same place all year long. A few birds have short migrations. They spend summer on the top of a mountain. When winter comes, they fly to the bottom of the mountain. Other birds travel a few hundred miles. But there are some birds that migrate thousands

Turtledoves

of miles. They fly all the way from the top of the world to the bottom and back again!

Time to do Activity 79 in the Activity Book!

How Do Birds Find Their Way?

We don't understand exactly how birds know the way to the same winter homes and summer nesting places each year. They seem to know which direction to go by looking at where the sun rises and sets each day. At night, they may use the stars the same way. Birds also seem to be able to tell direction by something inside them that senses where the North and South Poles are. They may know the place to stop by recognizing the land and water they see below them as they fly. Some birds might remember where to land by how the earth smells in a certain place.

God also made birds able to return in the spring to where they were born!

When a chick grows into an adult bird, God makes it able to migrate the first time all by itself. It doesn't need another bird to show it where to go. We don't know how birds find the way to their parents' winter home when they have never seen it before! God is amazing!

Time to do Activity 80 in the Activity Book!

Nesting

After a cold winter at the nesting place, the weather has warmed up enough to raise chicks. The birds come back from their winter home.

When it's time for the mother bird to lay her eggs, the father bird finds a good spot for a nest. He must find a place far enough away from other nests so there will be enough food for his family. The reason we hear so much bird song in the spring is that all the father birds are telling the other birds to stay away from their nesting areas.

The father's song does something else too: it attracts a female bird. Together they start building their nest. When the nest is finished, the mother bird begins laying eggs. She lays one egg a day until she has laid the number God has set for her kind of bird.

Blue tit at its nest

Sparrow and robin chicks take about two weeks to hatch from their eggs. Chicks are always hungry because they grow so quickly. Their parents are kept very busy feeding them! Only two weeks after hatching, chicks leave their nest and begin to fly. Baby birds do not have to be taught how to fly by their parents. That ability is something God has put in them! The father bird often teaches the young birds to feed themselves and watches over them as they practice flying. The mother bird is usually busy caring for another set of eggs.

Male cardinal feeding his babies

Prayer

Lord, thank You for making sure that birds can have babies! That way there will always be more birds. You have given birds' bodies all they need. You have also given them everything they need to know. It's amazing that Your wisdom tells them when to fly south and where to migrate each year. Since you care for the little birds, I know You love me too. Thank You! Amen.

Time to do Activity 81 in the Activity Book!

A dipper has caught a good meal in the stream.

CHAPTER 28

Beamer's Favorite Birds

And God blessed them, saying, "Be fruitful and multiply, and fill the waters in the seas, and let birds multiply on the earth." (Genesis 1:22)

Dippers

God made birds on the fifth day of creation—the same day He made fish. The Lord is very creative. He made some fish that fly and some birds that swim! Birds that swim underwater catch fish and other underwater creatures to eat. Penguins, loons, and cormorants are some of the fishing birds that can swim underwater.

Beamer's favorite swimming bird is called the **dipper**. Most swimming birds are not perching songbirds. But dippers are! Their song is a happy tune and chatter. Dippers are always bobbing up and down. Since they are usually standing at the edge of a stream, their bobbing dips them in and out of the water. They are always so busy watching for the best time to dive into the water that they never seem to be bothered by people nearby.

Dippers have extra oil on their feathers so water runs right off them. They have a lot of fluffy feathers next to their skin to keep them warm in icy-cold water. Dippers do not migrate to stay warm. The only reason they might move to a new place is if their stream freezes. Then they just go to another part of

the stream that isn't frozen. God made dippers content and happy to live in cold, wet places!

Dippers swim by flying under the water. They are looking for insects, worms, and fish. They will even move rocks with their feet to look for food! Dippers build their nests near streams—under bridges, high on stream banks, or even behind waterfalls!

Time to do Activity 82 in the Activity Book!

Hummers

Beamer also likes hummers! "Hummer" is the nickname for **hummingbirds**. Hummers are very small birds, but they show God's glory by the special things He made them able to do.

Your heart beats about 100 times in a minute. A hummingbird's heart can beat a thousand times a minute! Their

Another special thing about hummers is that they can see extra colors of light that people can't even imagine! That's why I like them!

Hummingbird with white tongue sticking out

little wings can flap more than 50 times a second. Hummers can move their wings differently than other birds. This makes them able to fly forward, backward, and even upside down. They can also fly in one place while they drink nectar from flowers.

Hummingbird on nest made of lichen and spiderwebs

Hummingbirds have long, thin beaks. Their beaks fit perfectly into the flowers God made for each of them. They drink nectar from these flowers and help to pollinate them. Some of these flowers could not be pollinated without hummingbirds.

Sweet flower nectar is the best way for hummingbirds to get the quick energy they need for their fast flying. God gave them special forked tongues that can quickly get the nectar they need. Hummingbirds also catch small insects to eat. Spiders are very important to hummingbirds. Hummers eat spiders and use spiderwebs to make their tiny nests.

Ruby-throated hummingbird

Even though hummers are warm-blooded, they can make their body temperature very low. They do this on cold nights to save energy so they won't have to eat as much. Some of them can get almost as cold as ice!

Most hummingbirds migrate each year. One kind even flies across 500 miles of ocean in only half a day. Before they travel, hummers stuff themselves until they have become almost too fat to fly. When they arrive, their bodies have used up their fat. They hope to find blooming flowers full of nectar right away!

Hummingbirds are very beautiful because of their bright colors. Their feathers have layers of little bubbles too tiny to see. The light going through these bubbles is separated and bent just like light in a raindrop when it helps make a rainbow. The bubbles are spaced perfectly to always make the same color on a certain kind of hummingbird. If the bubbles were just a tiny bit different, a color might be orange instead of red. But God is so amazing that He made a ruby-throated hummingbird to always have a throat the color of a ruby!

Beamer likes to shine on hummingbirds and bring out God's special colors for them.

Time to do Activity 83 in the Activity Book!

Honeyguides

For thousands of years, people have kept birds to use for meat, feathers, and eggs. These kinds of birds are not wild. They are not afraid of people. We call these kinds of useful birds **poultry**. Chickens, ducks, and turkeys are poultry.

Sometimes people catch wild birds as well. They teach or train these wild birds to help them. People have taught cormorant birds to fish for them. They have also used hawks to hunt other birds.

But of all the birds that are helpful to people, Beamer likes the **honeyguide** best! People don't own honeyguides, and they don't catch them. But people do use honeyguides to lead them to honey. The honeyguide is happy to help! It wants people to follow it to a bees' nest. This is because honeyguides eat wax honeycombs and baby bees. But they can't get to the honeycomb by themselves. They need people to pull the honeycomb out of the

bees' nest for them.

Sometimes the honeyguide asks people to follow it by chattering at them. It flies from tree to tree, chattering to make sure the people know where to follow. Other times, people give a special whistle to call a honeyguide. This whistle lets the bird know that they would like some honey and are ready to follow it. After the honeyguide and the people have found each other, the trip to the bees' nest begins. Once the bird has led the people to the nest, the people take out pieces of honeycomb and eat the honey. Then they leave behind some honeycomb and baby bees for the honeyguide to eat.

These are just three of the many interesting birds that God made. Each creature He made is special. Do you have a favorite animal? Take some time today to learn more about it. You will learn much about God's care and wisdom when you study His creatures.

Prayer

Lord, You are so wise in making each bird in an amazing way. They are all beautiful. They are each full of the good gifts You gave them. They use these gifts to do exactly what You created them to do. All of Your creation sings of Your glory! Amen.

Time to do Activity 84 in the Activity Book!

Worker bees tend to baby bees and honey in wax rooms of the hive.

UNIT 8

Animals That Make Milk

I'm excited to learn about this next group of animals God made: the beasts of the earth. Most beasts of the earth have hair. They give birth to cute little babies instead of laying eggs. And the mothers' bodies make milk for their babies!

Memory Verse

"For every beast of the forest is Mine,

And the cattle on a thousand hills.

I know all the birds of the mountains,

And the wild beasts of the field are Mine."

(Psalm 50:10,11)

Hymn Singing

Doxology

Praise God from whom all blessings flow;

Praise Him, all creatures here below;

Praise Him above, ye heavenly host:

Praise Father, Son, and Holy Ghost.

Amen.

You can listen to this hymn by searching for "Doxology for kids" on the internet.

Our memory verse for this unit shows that animals belong to God. Our hymn tells all people, creatures, and angels to praise God!

Bear nursing her cubs

CHAPTER 29

What Is a Mammal?

Mammals Make Milk

A **mammal** is an animal with certain gifts from God. Mammals almost always have hair. But some do not. Mammals almost always have four legs. But some do not. Mammals almost always give birth to live babies instead of laying eggs. But some do not.

Let's see how mammals are different from the animals we have learned about so far!

But there is one gift that God gave to all mammals: all mammal mothers make milk for their babies. The babies drink this milk by **nursing** from their mother.

Definition

Nursing is when a mother mammal is feeding her milk to her babies.

Even though we are not animals, God gave human mothers the ability to give milk to their babies too! Milk is a very special gift made for babies. God cares so much for babies that He made the perfect food for them. He also enjoys the praise their voices give Him. Jesus said:

"'Out of the mouth of babes and nursing infants
[God has] perfected praise.'"
(Matthew 21:16)

Just before a mother mammal gives birth, her body begins to make a special kind of milk. This first milk is different from the milk that will come later. The first milk has special things in it to help the new baby fight off germs. It also has things in it to help the baby's stomach and intestines start working. God is so wise to give newborn animals the start they need!

The milk that comes a little later will be all the baby animal needs until it can eat food like its parents. God has made each kind of animal's milk special. A baby walrus will get milk with a lot of fat in it. The fat helps the baby make its own layer of fat so it can stay warm in cold water. A baby camel will get milk with extra water in it to help it live in the dry desert. A baby whale gets milk as thick as toothpaste. Thin milk would wash away in the ocean before the baby whale could drink it!

Lioness feeding her cubs

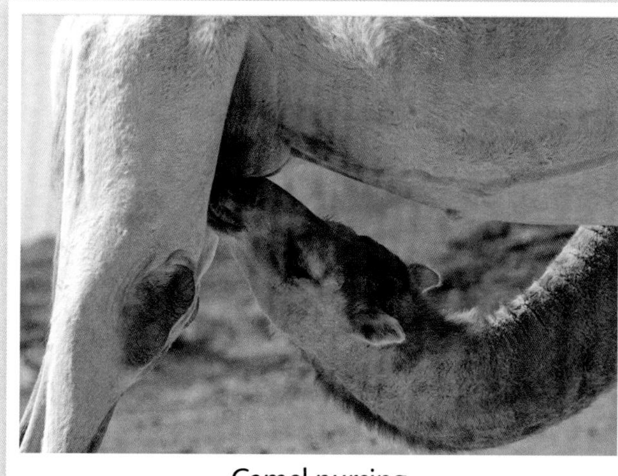

Camel nursing

Time to do Activity 85 in the Activity Book!

Mammals Have Hair

Mammals are warm blooded like birds are. They must eat a lot of food to get energy to keep their bodies warm. To help keep them warm, God has given most mammals hair. Most animal hair is called **fur**.

God made different kinds of hair for different jobs.

1. **Underfur** is short fine hair next to the skin. It helps trap air to keep the animal warm. Do you ever get little bumps on your skin when you are cold? These bumps are made by little muscles in your skin that are making the hair stick up. When animals are cold, they can make their underfur stick up to trap more air underneath.
2. **Guard hairs** are long, thick hairs that cover the underfur. They give the mammal its color. The color and pattern of an animal's fur helps it hide. Guard hairs are waterproof to help keep the mammal dry.
3. Some animals have **hollow hairs** to help keep them warm. Air is trapped inside each hair. Deer and elk grow hollow hairs in cold weather. They shed them in warm weather.

Deer in winter

Definition

A **hair follicle** is the place where a hair grows out of the skin. Oil also come out of hair follicles.

Whiskers help a mammal feel its way in the dark.

4. **Whiskers** are special stiff hairs that help an animal feel things. God put a lot of nerves at the place where the whiskers are attached to the body. Nerves sense things and send messages to the brain. Every time a whisker brushes up against something, the nerves send a message to the brain to let the animal know that something is there.

Porcupine with its quills raised

5. **Quills** or **spines** are hairs that are thick, stiff, and sharp. Porcupines and hedgehogs have spines for protection. They can use the muscles in their skin to make their spines stand up. This usually scares away an enemy.

Time to do Activity 86 in the Activity Book!

Mammals Are Born, Not Hatched

Mammals don't lay eggs. Instead, the baby grows inside its mother. Some mammals have only one baby at a time, like elephants. Sometimes more than one baby grows inside the mother at the same time. An opossum can have 30 babies at a time!

Baby mammals spend different amounts of time growing inside their mother. A baby African elephant grows for almost two years before being born. Baby mice only grow for 20 days inside their mother.

Kangaroos are an interesting mammal whose babies are born after only 34 days inside the mother. Newborn baby kangaroos are only about the size of a jellybean. Mother kangaroos have a special pouch made of skin where their babies can live after they are born. The newborn kangaroos climb up the outside

Fighting wolves

Bighorn sheep fighting

of their mother's tummy until they find the opening of her pouch. Then they crawl inside, find the place God made for milk to come out, and spend about six months growing inside their mother's pouch.

[God said,]
"Do you know the time when the wild mountain goats bear young? Or can you mark when the deer gives birth?"
(Job 39:1)

Newborn babies of hoofed animals can stand up and walk as soon as they are born. They have hair and can open their eyes right away. Newborn whales are able to swim as soon as they

are born. God knew they would need to go up for air right away! But most other animals need their mother's help when they are born. Newborn kittens can't see or hear. They can't keep their body temperature warm yet. They can't even empty waste from their bodies without the mother cat's help. Yet God makes each mother mammal know what her babies need.

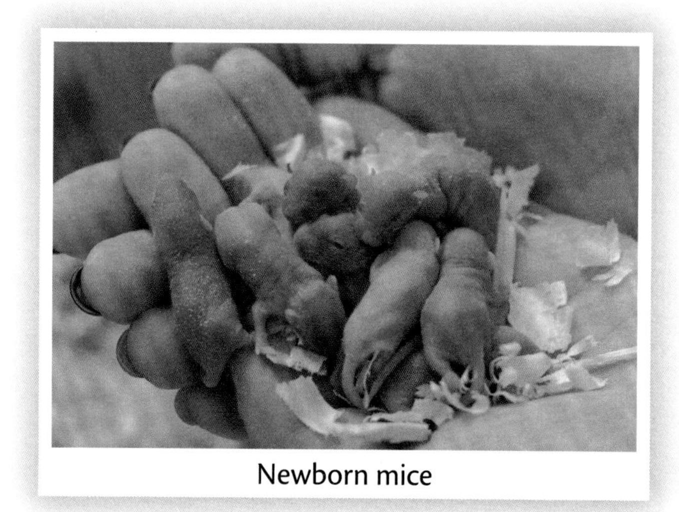
Newborn mice

Baby kangaroo in mother's pouch

There are actually two kinds of mammals that are not born as babies. Platypuses and echidnas are funny mammals that lay eggs. These animals look like silly mixtures of other animals. You may like to learn more about them on your own.

Dolphins

Newborn kitten

Prayer

Father, thank You for baby mammals that are born so cute and furry. We praise You for giving them milk from their mothers. You care for them and give them exactly what they need to start out their lives! Amen.

Time to do Activity 87 in the Activity Book!

CHAPTER 30

Hoofed Mammals

God Gave Gifts to Hoofed Mammals

The LORD God is my strength;
He will make my feet like deer's feet,
And He will make me walk on my high hills. (Habakkuk 3:19)

God made many kinds of mammals with **hooves**. Hoofed animals eat plants and must spend most of the day eating. God gave special gifts to these interesting animals:

1. God gave hooves to these mammals to protect their feet. Hooves also grab the ground and keep the animal from sliding when it runs or climbs. Different animals can have one, two, or more hooves on each foot.
2. God gave some hoofed

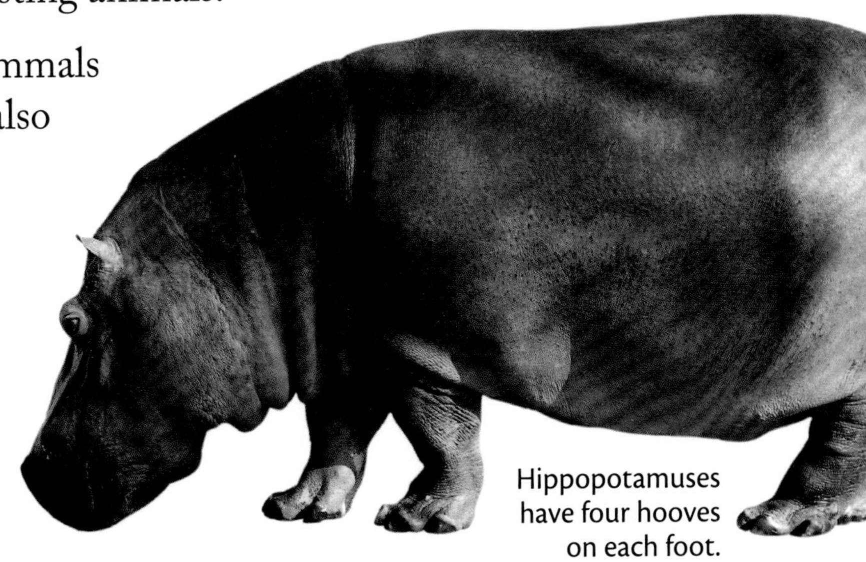

Hippopotamuses have four hooves on each foot.

Horse hooves

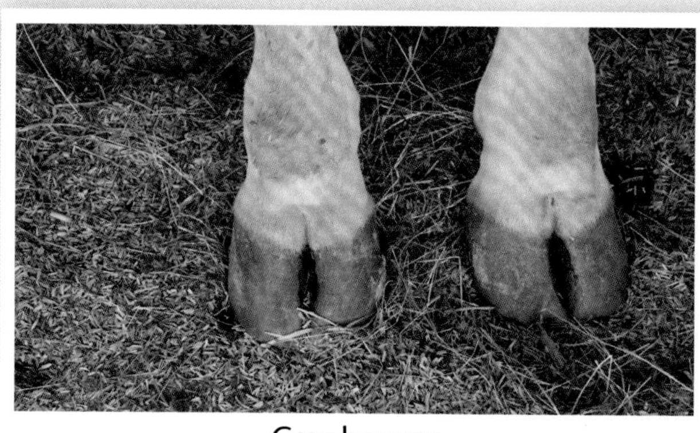
Cow hooves

Rhinoceroses have three hooves on each foot.

animals bodies made for running. An animal is fast if it can take a lot of steps quickly. A fast animal often has stretchy ligaments that help its legs spring back after its feet have left the ground. This helps it take its next step sooner. An animal is also fast if it can take long steps when it runs. God gave fast animals long legs to take long steps. Fast animals run on their toes. In fact, their hooves are at the ends of their toes. This gives them extra spring to take longer steps. They also have loose shoulders that let them reach farther forward with each step.

3. God gave many hoofed mammals hard things growing on their heads.

   - **Horns** usually grow near the top of an animal's head. Horns are made of bone with a covering like your fingernails. Both male and female animals have horns. They use their horns to fight or to protect themselves. Horns do not fall off each year. Instead, they keep growing. Cows, sheep, and buffalo are some animals that have horns.
   - **Antlers** are also near the top of an animal's head. They are made of bone and are covered with velvety skin for a short time. Antlers are usually branched.

Bighorn sheep with curled horns

Male elk with antlers

They fall off every year, and larger ones grow to take their place. Usually only males grow antlers. Antlers are used for fighting and protection. Deer and moose are some animals that have antlers.

- **Tusks** are large teeth that grow outside the mouths of some animals. These tusks are sometimes used for fighting. But mostly they are used to dig up food. God gave tusks to both males and females. Elephants, walruses, and hogs all have tusks.

Warthog

Time to do Activity 88 in the Activity Book!

Helpful Hoofed Mammals

Some hoofed animals are useful to people because they are **domestic** animals. Some domestic mammals are used for their meat, milk, or fur. Horses and oxen have been used for their strength, especially before modern machines were invented.

Definition

Domestic animals are kinds of animals that God created to be tamed and kept by people so they can be useful.

**[God said,] "Have you given the horse strength?
Have you clothed his neck with thunder?"** (Job 39:19)

Beamer's List of Domestic Mammals with Hooves

1. **Horses** are strong and fast. People can ride horses instead of walking. Horses can also be used to pull wagons or carriages.
2. **Cattle** can be used for their meat. Their meat is called **beef**. People drink cow's milk and use it to make cheese, yogurt, butter, and ice cream. The skin of cattle is used to make leather belts, purses, and furniture. Some cattle are used to pull heavy loads.

Horses

Cattle

Goat

Sheep

3. **Goats** are used for their milk and meat.
4. **Sheep** are useful for their meat, milk, and fur. Sheep fur is called **wool**. Clothes made of wool are very warm. Wool is warm even when it's wet!

Pigs

A llama train

5. **Pigs** are used for their meat. Their meat is called **pork**.
6. **Llamas** and **alpacas** are used to carry loads. Their long hair can be made into clothes.
7. **Camels** are very strong and can carry up to 900 pounds (400 kg). When God made camels, He gave them gifts to help them live in deserts. When sand

Camel

is blowing, camels can close their nostrils. Their eyes also have special protection. Each eye has three eyelids and two rows of eyelashes to keep out blowing sand. Camel humps can store 80 pounds (36 kg) of fat. They can live on this fat for weeks. They can drink 40 gallons (150 L) of water at a time! People sometimes drink camel milk, eat camel meat, burn camel waste for fuel, and make cloth out of camel hair.

Time to do Activity 89 in the Activity Book!

Wild Hoofed Mammals

Wild animals with hooves often live in groups called **herds**. Animals that stay in a herd are safer from enemies than if they were alone. This is because several animals can keep watch and warn the others that are busy eating. If an enemy does come, a herd will often gather closely together. They put the babies in the middle and protect them by using their hooves, antlers, or horns to hurt the enemy.

Beamer's List of Wild Mammals with Hooves

1. **Zebras** are a kind of wild horse. God gave them interesting black and white stripes. The stripes may confuse biting flies and keep them from landing on the zebra.[16]
2. **Bison**, **water buffalo**, **musk oxen**, and **yaks** are kinds of cattle that still live in wild herds. Some have also been domesticated for their meat, milk, and

Zebras

Water buffalo

leather. Water buffalo and yaks are used to carry heavy loads or pull plows. Some people raise bison for their meat, but they are not tame like cattle. They can be dangerous and need sturdy fences to keep them in.

3. **Mountain goats** and **ibexes** are wild goats. They are very good at climbing. God has given them special pads in the middle of their hooves that help them grab and hold onto rocks.
4. **Bighorn sheep** are a kind of wild sheep. Males have large curled horns that they use for fighting other sheep.
5. **Peccaries, warthogs, babirusas** and **wild boars** are kinds of wild pigs. They all have tusks.

Baby mountain goat

Alpine ibex

Running pronghorn

Impala, an African antelope

6. **Antelope** and **pronghorns** are graceful, fast runners. They have horns on their heads.

7. **Deer**, **elk**, **moose**, and **caribou** (or reindeer) have antlers on their heads. Usually only males have antlers. But female caribou also have them. These animals lose their antlers and grow new ones each year.

8. **Giraffes** are the tallest mammals on Earth. Their legs are as tall as a man! Even though giraffes have such long necks, they cannot drink without spreading their front legs apart. You need about 10 hours of sleep a day. Giraffes only need 5-30 minutes, and they sleep standing up!

Roe deer

Moose

Giraffe spreading its legs to reach water

Adult rhinos can weigh more than a ton (1000 kg).

9. **Rhinoceroses** have one or two thick horns on their noses. Their skin is very thick and looks like armor.

10. **Hippopotamuses** are very large mammals that like to spend their time in the water. Their babies even nurse underwater! Hippos swim well because they have webbing between their four toes. Each toe has its own hoof. The only hair on a hippo is around its mouth and on the tip of its tail.

Prayer

Lord, You have made pudgy pigs and giant giraffes. And You made them both with hooves! We don't know how You thought of so many different animals. Hoofed animals remind us that You care for us because You made so many of them to be useful. Thank You! Amen.

Time to do Activity 90 in the Activity Book!

Grizzly bear

CHAPTER 31

Paws and Claws

Mammal paws have claws attached to them. Claws can be scary because they are sharp and dangerous.

As a boy, David protected his father's sheep from the paws of a lion and a bear. But God was protecting David at the same time. When Goliath, the wicked, giant Philistine, called for someone to fight him, David said he would do it. He knew God would protect him again!

David said, "The LORD, who delivered me from the paw of the lion and from the paw of the bear, He will deliver me from the hand of this Philistine." (1 Samuel 17:37)

Male lion

Tigers have canine teeth.

What are Paws?

Paws are a type of feet given to some animals. Mammals that do not have hooves often have paws. Paws do not have fingernails or toenails. Paws have **claws**. Claws are sharp tools that are useful for many things. They can be used to catch other animals. They can also be used to fight against an enemy. Claws can help an animal climb a tree or dig in the dirt. They can be used to scratch an itch!

Dog claws always show. Cat claws only show when the cat wants to use them. The rest of the time they are hidden in its fur. When the cat wants to use them, it stretches its foot like we do when we point our toes. That stretch makes the claws come out.

Animals with paws have pads on the bottom of their feet. Each paw has one large pad and four or five smaller pads at the toes. Claws are attached to the toe pads. Paw pads are often dark-skinned and usually have no hair on them. Fat inside the pads helps to cushion the animal's steps.

Many animals with paws eat other animals. We call them **meat eaters**. Meat eaters often have four teeth that are longer and sharper than their other teeth. These are called **canine**

teeth. These teeth help the animal grab and tear meat. Meat eaters also have very strong mouths. Most meat eaters can eat plants as well.

Time to do Activity 91 in the Activity Book!

Mammals with Paws, Claws, and Canine Teeth

Beamer's List of Mammals That Have Paws and Eat Meat

1. Dogs can be wild or domestic. When dogs are domestic, they are usually our pets. Sometimes people also use domestic dogs to herd sheep, guard homes, pull sleds, or catch criminals. Some dogs can help blind people walk safely. They are called **Seeing Eye dogs**.

   Dogs have a very good sense of smell. Some can use their noses to find lost children. The dog will smell some of the child's clothes and then follow the same smell to where the child is. Some dogs are trained to smell for bedbugs in houses. Others can smell if a person has a certain sickness.

Seeing Eye (guide) dog

   Wolves, **foxes**, **coyotes**, and **jackals** are wild dogs. Wolves usually hunt large, hoofed animals. They hunt in groups called **packs** so they can attack an animal from all sides.

   If a coyote is by itself, it will hunt small animals. But if it is with a pack of other coyotes, it will hunt larger animals.

Foxes hunt alone. They eat plants when they can. But in winter they must hunt animals to eat. Like all dogs, foxes have good ears. They can even hear animals burrowing underground! They dig them up and catch them.

Jackals hunt alone or in pairs. They eat animals and plants.

Fox jumping for a bird

2. **Cats** can be wild or domestic. Domestic cats are pets. They are sometimes helpful by catching mice. Some of the world's wild cats are **lions**, **tigers**, **cougars**, **lynxes**, **bobcats**, **leopards**, **jaguars**, and **cheetahs**. Most wild cats live and hunt alone. Lions are the only cats that live and hunt in groups. The females in a lion group are all from the same family. They are mothers, grandmothers, daughters, and sisters to each other. The females hunt together. The male lions in each group are not part of the females' family. The males hunt alone.

Jackal

3. **Bears** eat both plants and animals. Most bears hibernate during the winter. Some do not hibernate because they live in warm climates

The cheetah is the fastest land animal.

with lots of food. Male **polar bears** also do not hibernate because they can hunt for seals under the ice all winter. Female polar bears do hibernate. They give birth to their two cubs during that time.

Polar bear mother and cubs

4. **Raccoons** eat both plants and animals. Their front paws look like little hands with claws instead of fingernails. They can feel things very well with their front paws. This helps them catch creatures in muddy water. They even have thumbs that allow them to use their paws like we use our hands. Raccoons are mostly active at night. After storing up body fat, they spend much of the winter in their dens. But they don't hibernate.

5. **Skunks** are known for their bad smell. They have two places under their tail that can make a strong-smelling liquid. If they are attacked, they can spray the enemy with their stinky liquid to make it run away. Skunks eat both plants and animals.

This young bobcat hasn't learned about skunks yet.

Sea otter using its tummy for a table

6. **Weasels**, **badgers**, **minks**, **ermines**, and **otters** are small animals. They have long bodies, short legs, little round ears, and thick fur. People sometimes use the fur of mink, ermines, and otters for coats. Most of these mammals live alone. They are active at night and do not hibernate. They eat only meat. A weasel can kill an animal larger than itself!

Sea otters eat sea creatures that have shells. The sea otter floats on its back and uses its tummy like a table. It puts a rock on its tummy and bangs the sea creature's shell on the rock to crack it. Then it eats the soft insides of the creature.

Time to do Activity 92 in the Activity Book!

Mammals with Paws, Claws, and Teeth That Keep Growing

Rodents and **rabbits** have paws with claws. Rabbit paws do not have pads. Instead, they have thick hair to cushion their feet.

Rodents and rabbits do not have canine teeth. They have big teeth in the front of their mouths called **incisors**. They also have chewing teeth in the back. In between these two kinds of teeth, there is a big space with no teeth at all. Rodent and rabbit incisors are always growing longer. They must chew on tough food to keep these teeth from getting too big. If they don't eat tough food, their teeth can grow so long that they can't use their mouths to eat at all!

A marmot shows its four incisors.

There are more rodents and rabbits in the world than any other mammals. They have a lot of babies very often. God made it that way because they are food for meat-eating animals.

Some of the rodents God made are **mice**, **rats**, **squirrels**, **beavers**, **porcupines**, and **capybaras**. Rodents can be pests because they spread disease, eat our food, and destroy our things by chewing on them. Rodents have four incisors—two on top and

Mouse

Squirrel

Beavers eat wood.

Porcupine

Capybara

two on the bottom. They are always gnawing on things to keep their incisors short. This can cause a mess if they get into your house!

In the rabbit group of mammals, there are **rabbits**, **hares**, and **pikas**. These animals all have six incisors. They have the same four incisors that rodents have, but they also have two extra ones behind their top incisors. These extra teeth are called **peg teeth**. They are round and flat on top like pegs. The bottom incisors cut food against the peg teeth.

Hares are larger than rabbits. When baby hares are born, they have hair and can see. They can live on their own when they are only an hour old.

Baby rabbits are born blind, have no hair, and must be taken care of by their mother until they get older.

Rabbit

Pika

Pikas live in rocky places in the cool mountains. They have short, round ears. They don't hibernate, so they must save enough food to last them through the winter. Pikas cut down plants with their teeth and spread them on rocks to dry. Then they gather the dried plants into big haystacks they keep under the rocks.

Prayer

Thank You, God, for furry animals with paws. Thank You for pets. We're glad You gave some animals claws to use as tools. We're also glad You gave mammals different kinds of teeth for the food they eat. Amen.

Time to do Activity 93 in the Activity Book!

Elephant trunks are very strong.

CHAPTER 32

Feet, Flippers, and Wings

Some Feet Are Called Feet

Some mammals have feet that are not paws or hooves. Mammal feet with flat nails are called **feet**.

Elephants seem a lot like hippopotamuses and rhinoceroses. They are big, gray, and only have a little hair. But they don't have hooves. Instead, their feet are called feet. These feet have flat nails like your toenails on each toe. Nails do not wrap around a mammal's toes like hooves do.

Elephant trunk and foot

Many elephants also have tusks. They use their tusks for digging, lifting, gathering food, and protecting themselves. The tusks protect the elephant's trunk as well. An elephant uses its trunk for drinking, breathing, and eating. The trunk can also do small jobs for the elephant because it

God joined an elephant's nose and its upper lip together to make its trunk!

Spider monkey

Orangutans are apes.

has things like fingers at the end of it.

Monkeys, apes, and lemurs also have nails instead of claws or hooves. God gave them thumbs like you have so they can grab things. He also gave them big toes that are like thumbs so they can grab with their feet too! They are very good at traveling through the forest. They use their hands and feet to swing through the trees.

Monkeys have tails. Some monkeys can grab tree branches with their tails and swing with them.

Apes are bigger than monkeys and do not have tails. They usually have no hair on their faces, and their arms are longer than their legs. **Orangutans,**

gorillas, and **chimpanzees** are apes.

Lemurs have tails, but they do not grab with them. Lemurs are active at night, so God gave them big eyes with a **tapetum**. The tapetum helps their eyes see better in the dark. Lemurs always have wet noses!

Ring-tailed lemurs

Time to do Activity 94 in the Activity Book!

Mammals with Flippers

God made some mammals to live in the ocean! They have **flippers** instead of legs. Flippers are flat with finger bones or toe bones inside. All sea mammals are meat eaters. They have **blubber** to help keep them warm.

> **Definition**
>
> **Blubber** is a thick layer of fat under the skin of sea mammals.

Whales never come on land. They were probably made on the fifth day of creation along with fish. Since whales are mammals, they have to bring air into their lungs. But whales cannot breathe through their mouths like most mammals. Their mouths are only for eating. They do not have nostrils like you do. They breathe through **blowholes** on top of their heads. When they come up for air, they first blow out the old air. We can sometimes see this moist air spraying out.

Whales only have front flippers. They use them for steering. Their tails are called **flukes**. Whale flukes move up and down instead of sideways as fish tails do.

Killer whale blowing moist air

Definition

Blowholes are the nostrils on top of a whale's head that it breathes through.

There are two kinds of whales: toothed whales and baleen whales.

Most **toothed whales** are smaller than baleen whales. They use their teeth to catch and chew the animals they eat. They have only one blowhole. **Sperm whales**, **killer whales**, **beluga whales**, **narwhals**, **dolphins**, and **porpoises** are all toothed whales.

Baleen whales are usually larger than toothed whales. They have two blowholes. Instead of teeth, they have **baleen**. They hunt for swarms of fish, krill, and other small creatures to eat. When they find them, baleen whales swim toward the creatures with their mouths open. They use their baleen as a net to capture as much as they can. Then they push only the water out of their mouths through the baleen and swallow the food whole. The **blue whale** is a baleen whale. It is

Male narwhals usually have a tusk that is a canine tooth.

Dolphin

Definition

Baleen is a tough fringe that hangs down from the upper jaws of whales without teeth. It is used to strain small ocean animals out of seawater.

Baleen on gray whale

the largest animal that has ever lived. Other baleen whales include the **bowhead**, **right**, **gray**, and **humpback whales**.

Right whale

Seals are sea mammals with four flippers. Seals have whiskers on their noses that help them feel even small motions. They are good swimmers but have trouble moving on land. They come on land to raise babies or to escape from enemies. There are three kinds of seals:

1. **Eared seals** have ear flaps that you can see. They swim by pushing themselves through the water with their long front flippers. They use their back flippers for steering. On land, they pull their back flippers under their body and use their front and back flippers like legs. Male eared seals are much bigger than the females. They can be very loud as they tell the other males to stay away. **Sea lions** and **fur seals** are eared seals.

2. **Earless seals** really do have ears, but they don't have ear flaps. They have short necks and short front flippers. They cannot pull their back flippers under them to help them move on land. On land, earless seals move like caterpillars by throwing the front of their body forward and letting the back follow. **Monk seals**, **harbor seals**, **leopard seals**, and **elephant seals** are all earless seals.

Leopard seal

Fur seal

3. **Walruses** are a large kind of seal. They don't have ear flaps. Instead of canine teeth, they have long tusks. They use their tusks to pull shelled creatures off of underwater rocks to eat.

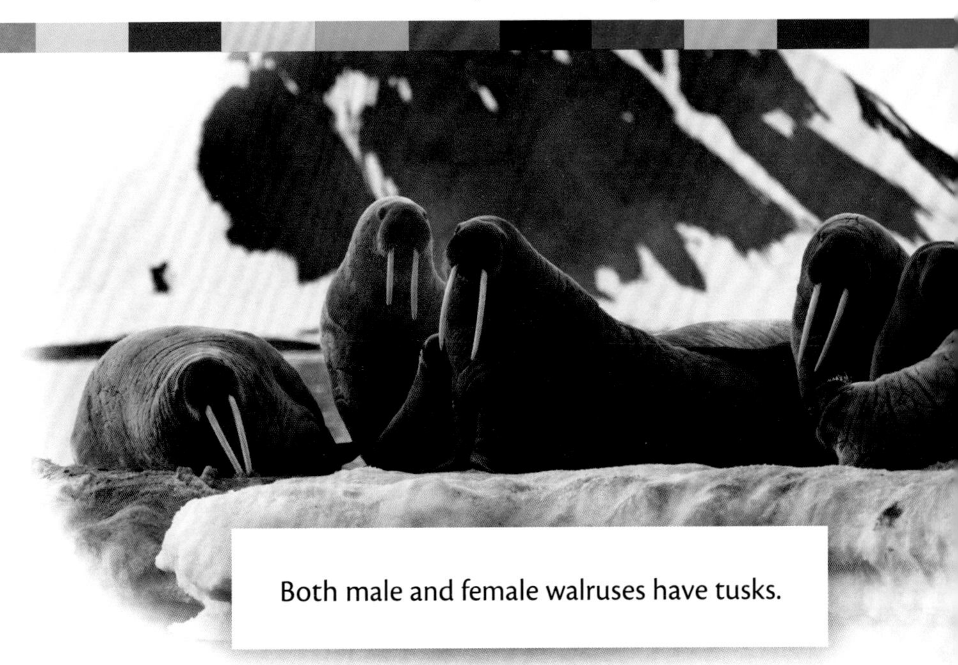
Both male and female walruses have tusks.

Time to do Activity 95 in the Activity Book!

Winged Mammals

Bats are the only flying mammal in the world. There are a few mammals that can glide downward from the trees, but bats have flapping wings. A bat's wings are made of skin stretched between its fingers. People used to think that bats were flying mice. But if you look at a bat's teeth, you will see that they are not rodent teeth!

Bat teeth

Bats are important for pollinating bananas, mangoes, and avocados. They also help keep the number of insects low. A bat can eat about a thousand mosquitoes an hour. Bats are active at night. How do they fly and catch food in the dark? God has given them **echolocation**.

Bats send out clicks from their mouths and gather the sound's echo in their big ears. A bat can find a tiny fruit fly in the dark 100 feet (30 m) away!

Echolocation is a way some animals use to sense things when they cannot see. They send out sounds and listen for the echoes to come back.

"I will bring the blind by a way they did not know;
I will lead them in paths they have not known.
I will make darkness light before them,
And crooked places straight.
These things I will do for them,
And not forsake them."
(Isaiah 42:16)

God is so kind to give bats a way to find their food in the darkness. In this verse, He promises help for people living in darkness!

Prayer

Lord, You have made mammals that can swim, fly, and walk. You have made so many of them with interesting legs and arms that are just right for getting around the neighborhoods they live in. You have even made bats to fly and eat all night long. Thank You! Amen.

Time to do Activity 96 in the Activity Book!

UNIT 9

You Are Wonderfully Made

Hasn't it been great to learn how special each part of God's creation is? Now we are going to learn about people. People are *very* special to God. He made people in His image.

God made people able to talk, to think about complicated things, and to know Him. He made people to love, worship, and serve Him. And the human body is amazing!

Here is a verse to memorize and a hymn to sing as we learn about the human body!

Hymn Singing

Father, We Thank Thee

Father, we thank thee for the night,
And for the pleasant morning light;
For rest and food and loving care,
And all that makes the world so fair.

Help us to do the things we should,
To be to others kind and good;
In all we do, in work or play,
To love Thee better day by day.

You can listen to this hymn by searching for "'Father, We Thank Thee' Cedarmont Kids" on the internet.

Memory Verse

I will praise You, for I am fearfully and wonderfully made.

(Psalm 139:14)

CHAPTER 33

The Way You Are

And the LORD God formed man of the dust of the ground, and breathed into his nostrils the breath of life; and man became a living being. (Genesis 2:7)

You are the way you are because God made you that way! Let's learn about how special God made humans to be.

God made people after He made everything else. He made us different from animals. God has given people special gifts:

- We are made in God's image.
- We are created with a soul that will live forever.
- We are able to tell the difference between right and wrong.
- We can talk, sing, and praise God.
- We can think about complicated things.
- We can use things from God's world to create other things.
- We are made to rule over animals.
- We can know God!

Time to do Activity 97 in the Activity Book!

People Are Different from Each Other

God has made *each person* different from anyone else. Some people have curly hair; some have straight hair. We each have our own eye color and skin color. Our voices sound different. We have different ways of walking. Some children are boys, and some are girls. There has never been anyone exactly like you! There never will be anyone like you!

God has made each person special. He cares for each of us. He listens to each person's prayers. He has a plan for each of our lives. The Bible says God wrote about our days before we were even born.

**In Your book they were all written,
The days fashioned for me,
When as yet there were none of them. (Psalm 139:16)**

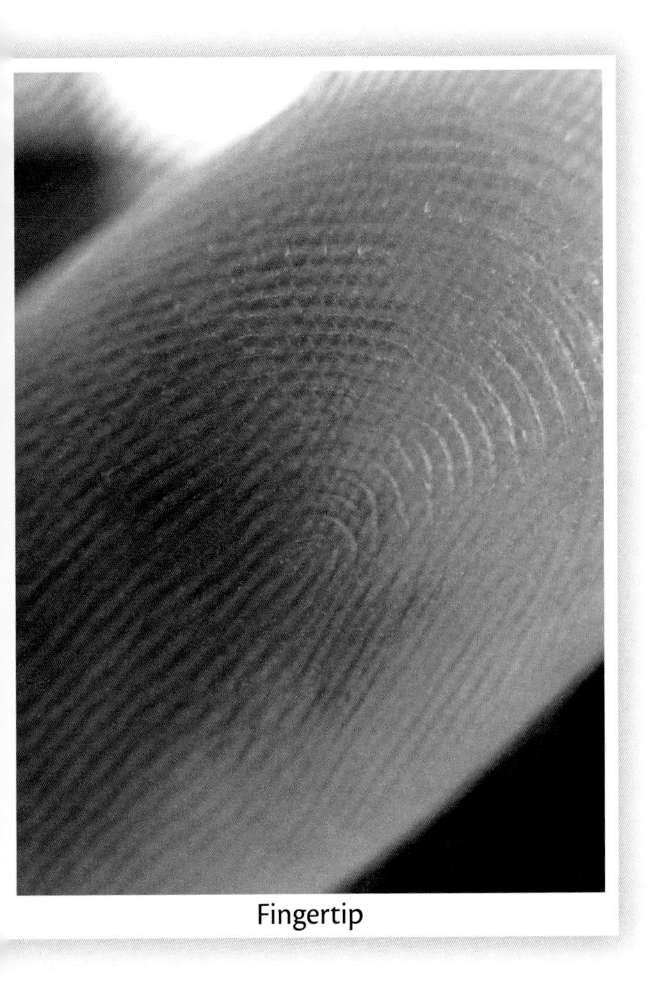
Fingertip

Look at the tips of your fingers. Do you see little lines and swirls on them? Each person's fingertips are different. Even twins that look alike have different designs on their fingertips.

God gave you these designs in a very special way. Fingertip designs are made when babies are still inside their mothers. The skin on their fingertips starts to grow faster underneath than it does on top. This makes the skin buckle. It buckles differently for each baby because of the way the baby touches things inside its mother. As the skin buckles, it forms lines and swirls on the fingertips. The design on babies' fingertips is permanent. It will stay the same their whole lives.

It is easy to see fingertip designs by making a print of them. We can press our fingertips onto an ink pad and then onto a piece of paper. The paper will show a print of our fingertip design. By doing this, we have made **fingerprints**.

Fingerprint

You Began inside Your Mother

Living things are made of smaller living things called **cells**. We can't see cells unless we use a microscope. Cells are like little bodies. Each cell has something like a brain that makes it work. Each cell has a way to take in food and put out waste. Cells also have ways to make more of themselves. They don't have babies. But they break themselves in half, and each half grows into a new cell. Each cell knows what to do because it has instructions inside it. We call these instructions **DNA**.

Time to do Activity 98 in the Activity Book!

For You formed my inward parts; You covered me in my mother's womb. (Psalm 139:13)

Do you know who wrote the instructions to make you? God did!

You started out as a tiny little cell. But you didn't stay little for long! Soon your one cell became two. Then you became four cells, then eight, then sixteen, and you have been growing ever since! God gave your mother a special place called a **womb** for you to grow inside. About five days after God created you, you snuggled into the side of your mother's womb and soon became attached there.

In the next few weeks, your cells started to become different from each other.

Definition

DNA is the set of instructions God put inside each cell. The instructions tell the cell what to do and how to do it. DNA looks like a long, twisted ladder.

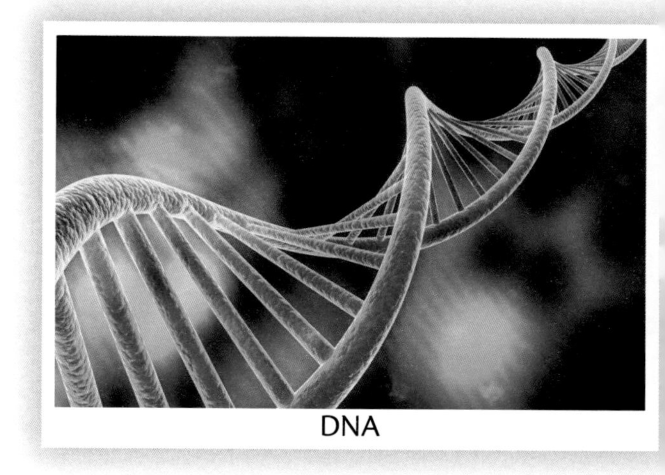

DNA

as they broke in half to make new ones. Some became brain cells. Some became bone or muscle cells. Your eyes formed. You grew skin. But right from the start, you were a real person with all the instructions to become what you are now.

For nine months you grew inside your mother. Food and oxygen went from your mother's blood to your blood through a long cord. Waste and carbon dioxide went out of your blood through the same cord. The cord was attached to the inside of your mother's womb and went to your belly. You got everything you needed through this tube. Then, at the perfect time, you were born. Happy birthday!

As soon as you were born, you needed to start breathing air. You needed milk for your food. Your body had to start taking energy and vitamins from the milk. You were too little to care for yourself, but God cared for you. Just like when you were a little cell inside your mother's womb, He cared for you, and He still cares for you!

A baby's monthly development in the womb

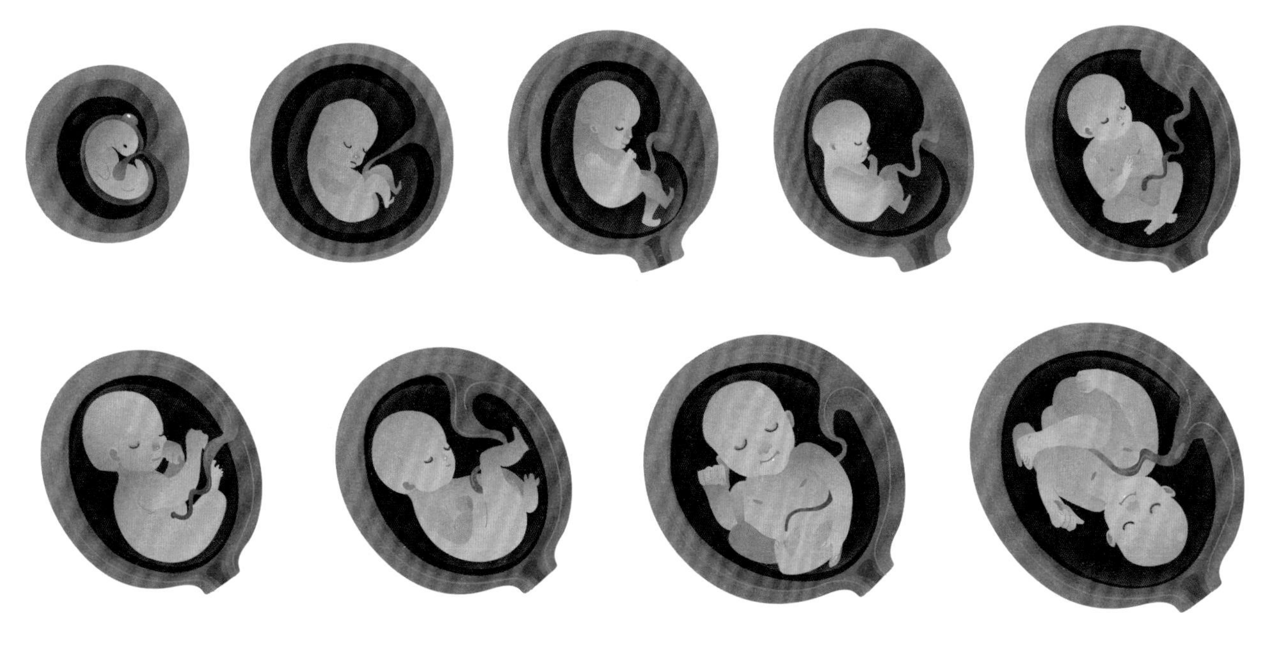

A clip is attached to a newborn's special cord before he is separated from it.

Prayer

We praise You, Lord! People are so different from anything else You created. You made each person special. You made us to love, worship, and serve You. We can't understand how You made us. We see Your greatness, Lord, by how complicated we are. Thank You that You made me. Thank You for putting me in my mother's womb. Thank You that I am who I am. Amen.

Time to do Activity 99 in the Activity Book!

CHAPTER 34

How You Know What You Know

God created our bodies to work with the world around us. Our senses gather information from the world. Then our brains use that information to learn and to know. Let's learn how this works.

[God asks,] "Who has put wisdom in the mind?
Or who has given understanding to the heart?"
(Job 38:36)

The answer to God's question is, "God"! He is the One who made you able to know and understand things!

Your Brain

Your **brain** is inside your head. It's made of different kinds of cells. The main kind of brain cell is called a **neuron**. Neurons get messages from every part of your body. They can also send messages to anywhere in your body. God gave you neurons so you can feel, move, think, and understand the world around you.

A neuron has a main body. It also has little branches sticking out of its main body to collect information. And it has a long, branching trunk to send information where it needs to go. Messages travel through each neuron using electricity. Neurons have spaces between them. Messages must jump across these

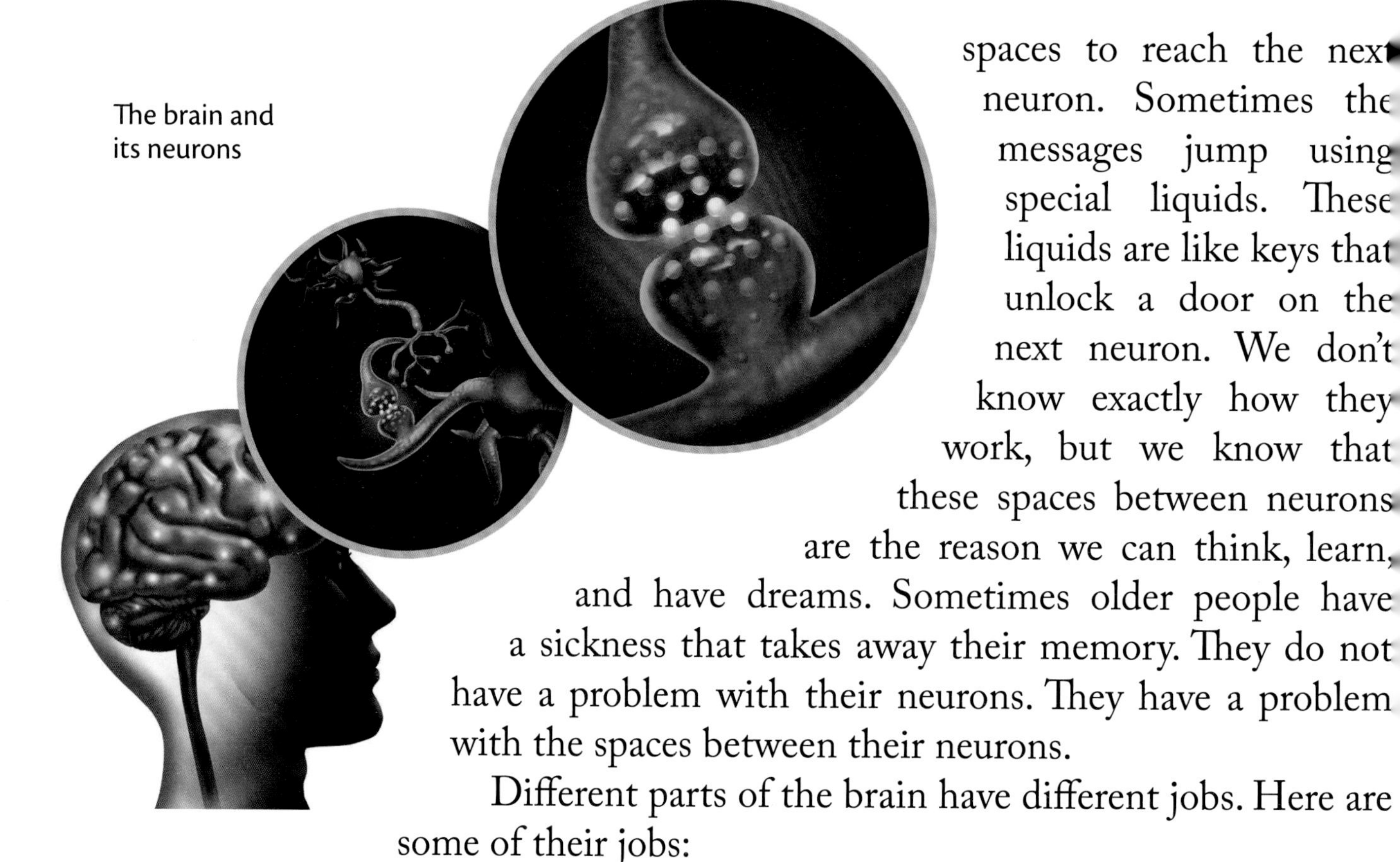
The brain and its neurons

spaces to reach the next neuron. Sometimes the messages jump using special liquids. These liquids are like keys that unlock a door on the next neuron. We don't know exactly how they work, but we know that these spaces between neurons are the reason we can think, learn, and have dreams. Sometimes older people have a sickness that takes away their memory. They do not have a problem with their neurons. They have a problem with the spaces between their neurons.

Different parts of the brain have different jobs. Here are some of their jobs:

Learning and talking

Knowing what position your body is in, knowing left and right

Seeing

Hearing

Balance, controlling your muscles

Time to do Activity 100 in the Activity Book!

Knowing Your World through Hearing and Touch

We have five senses: seeing, hearing, touch, taste, and smell. We have already learned about seeing in Chapter 4. Now we will learn about the other senses!

How We Use Sound

Sound is an energy we can't see. Sound travels in waves. But sound is different than light. Sound works by making things **vibrate**. Sound is able to travel through something solid like a wall because it makes the wall vibrate. Light cannot travel through a wall. Sound travels a million times slower than light. That's why you can see a flash of lightning before you hear its thunder.

Definition

Vibrate means to move back and forth very quickly. You can watch a rubber band vibrate by plucking it while it's stretched tightly.

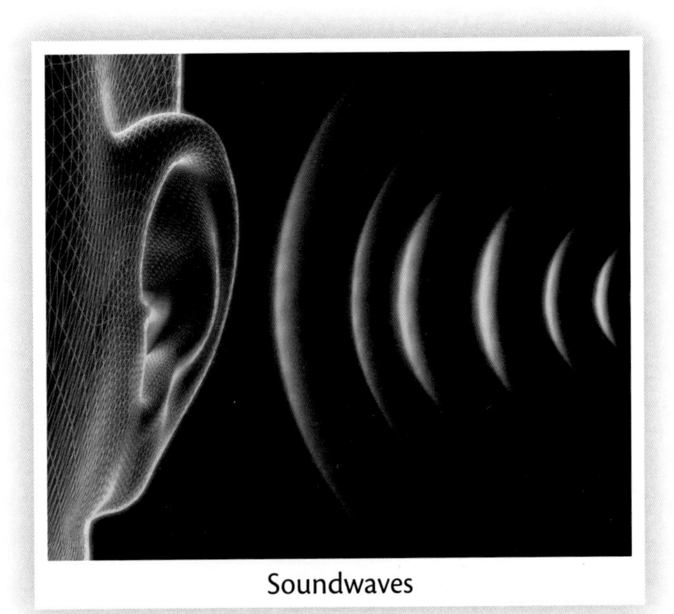

Soundwaves

Your ears can hear because God made them work with the vibrations of sound. When sound waves come into your ear, they first hit your **eardrum**.

The vibrations make the eardrum move. There are three tiny bones behind the eardrum. The movement of the eardrum makes the three bones vibrate. These vibrations travel farther into the ear. They go into a curled tube that looks like a snail. It has neurons in it that change the vibrations into information about the sound. This information can now travel through a special **nerve** to get to the brain. The brain makes us understand the information.

Definition

The **eardrum** is a thin piece of skin layered with something tough to hold it stiff. It is stretched tightly inside the ear so it can catch sound vibrations from the air.

A **nerve** is a group of neurons.

Inside the Ear

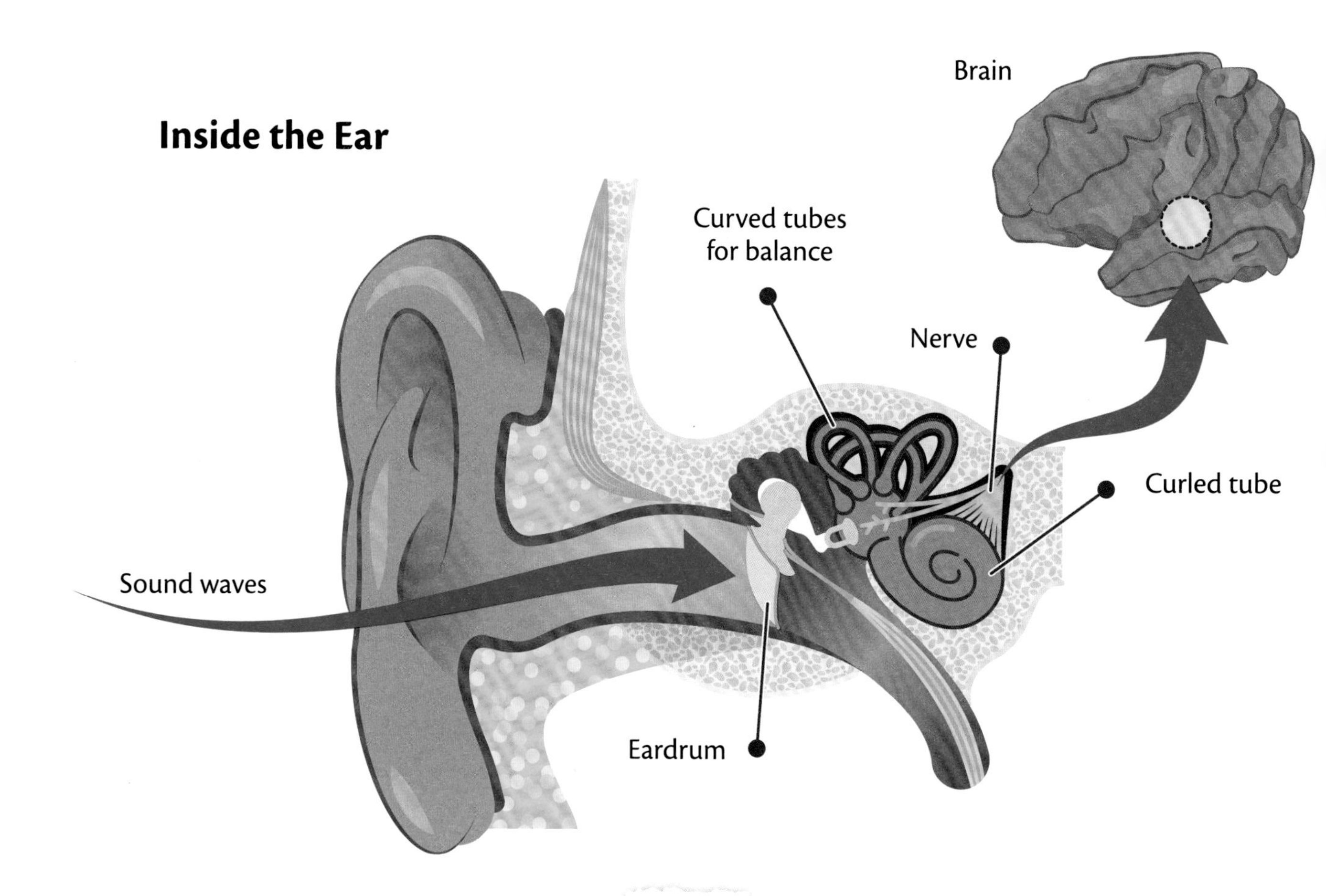

You also have three curved tubes in your ears that are filled with liquid. When you twirl or do a somersault, the liquid sloshes around. The tubes can feel what is happening to the liquid. Nerves in the tubes send that information to your brain. Your brain uses this information to help you keep your balance no matter what position your body is in.

How Touch Works

When we want to feel something, we usually use our fingers to touch it. The skin on our fingers has a lot of nerves in it. We have nerves in the skin all over our bodies. We have nerves for feeling inside our bodies too. Your stomach has nerves to tell you when you're hungry. If you stub your toe, you feel pain because of nerves inside your toe.

You have four kinds of nerves in your skin. These nerves can sense four kinds of touch. They send their information to the brain. The brain helps us understand it.

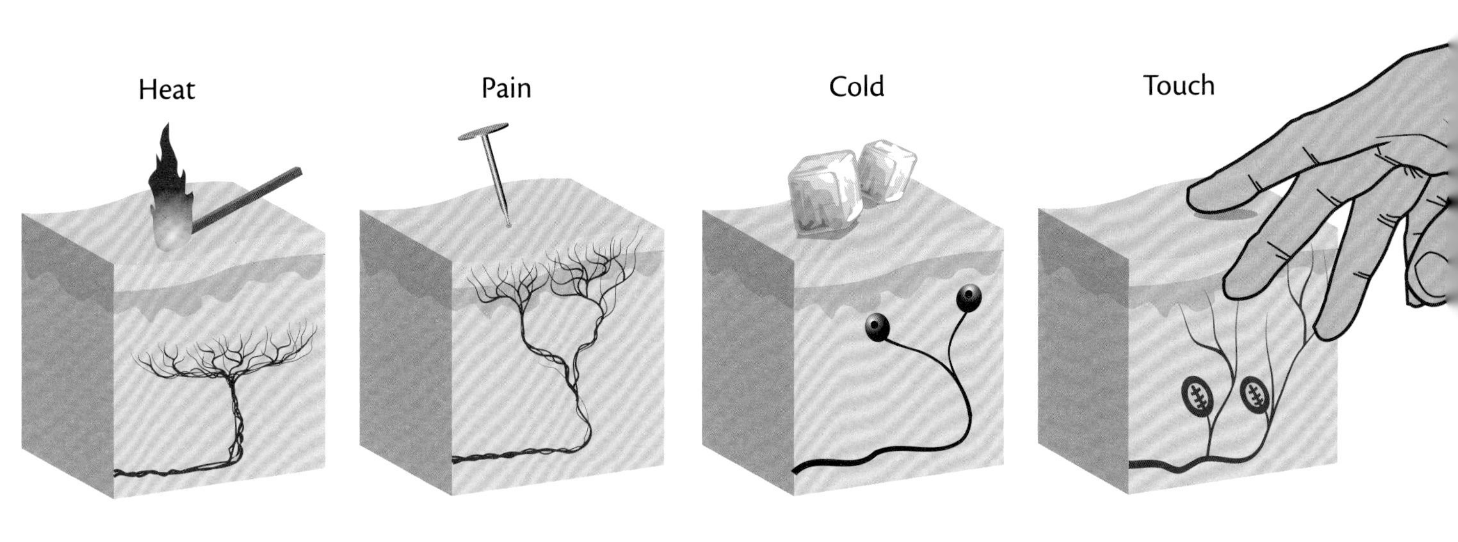

Time to do Activity 101 in the Activity Book!

Knowing Your World through Smell and Taste

Your senses of smell and taste work closely together. Your nose and mouth are almost connected. They both open into your throat.

How Smell Works

Smells come in through your nostrils as you breathe. The air passes a special patch of smelling cells high inside your **nose**. These cells are neurons that God designed for smelling. Each neuron can sense a certain kind of smell. Each neuron sends its information to your brain. The different information from several neurons helps your brain figure out what you are smelling.

Since your nose and mouth are connected inside, you can smell your food as you chew it. Smelling your food this way helps you enjoy it. You have probably noticed that when your nose is stuffy, food doesn't taste very good. That's because the smell of the food you are chewing can't get to the patch of smelling cells.

How Taste Works

Tasting is more than just smelling the food you are chewing. You also have special cells that send other information to your brain about what you are eating. These tasting cells are found in the little bumps on your tongue. These little bumps are called **tastebuds**. Tasting cells are also found on the roof of your mouth and in

your throat. There are five different kinds of tasting cells that taste five different tastes: sweet, salty, bitter, sour, or savory. Tasting cells are not neurons, but they give their information to neurons that send it to your brain.

Prayer

Dear Lord, thank You so much for giving us our brains and senses so we can know about the world around us. We enjoy seeing Your beautiful earth and sky lit by Your light. We like to hear music and our loved ones' voices. We can feel cold water and warm sunlight. You gave us fragrant flowers and fresh air to smell. You show Your love by making the food we need taste good. There is no end to the way we can use our five senses to enjoy Your blessings. I praise You! Amen.

Time to do Activity 102 in the Activity Book!

CHAPTER 35

Your Body Works Automatically

God made many parts of your body work automatically. This means that parts of your body work without you purposely making them work. Your body sweats when it knows you're too warm. If your body knows it doesn't have enough water, it will make you feel thirsty. Your body cleans out dead cells and grows new ones. Your eyes are always seeing and sending messages to your brain and blinking without being told. Here are three more jobs your body does automatically.

You Grow

And the child Samuel grew on, and was in favor both with the LORD, and also with men.
(1 Samuel 2:26 KJV)

Your body automatically started growing when God created it inside your mother. If you are a girl, you will keep growing taller until you are about 14 or 15 years old. If you are a boy, you will stop getting taller at about age 16. After that, your bones will get thicker and stronger for about 10 more years. The only bones that will never grow are the

three bones in your ear. They stay the same size from birth through the rest of your life.

Your muscles will automatically keep growing a few more years after you stop getting taller. As an adult, your muscles will not automatically grow, but they can get bigger with exercise.

Can you think of any parts of your body that never stop growing? They are things we need to cut now and then. The only things in your body that never stop growing are your fingernails, toenails, and hair.

Time to do Activity 103 in the Activity Book!

Your Blood Travels

God has given us our **blood** because we could not live without it. Blood is not just red-colored water. It is an amazing liquid. Let's learn what blood is made of and what it does. Blood is made of four things:

Red blood cells are the reason blood looks red. One drop of blood has millions of red blood cells in it. Every cell in your body needs **oxygen** to make energy. Red blood cells are able to absorb oxygen from your lungs as you breathe. They take the oxygen to your body's cells. Your body's cells need to get rid of **carbon dioxide**. Red blood cells are able to absorb carbon dioxide and take it to the lungs so it can leave your body as you breathe.

White blood cells are very busy. They rush to attack germs in your body. Some white blood cells kill germs by eating them. Others kill germs with poison. One kind of white blood cell puts a mark on germs so other white blood cells will kill them. Another kind learns how to recognize a certain type of germ so it can fight it more quickly the next time that type of germ attacks.

Platelets are pieces of cells that join together to plug up a place that is bleeding. Scabs are made of platelets.

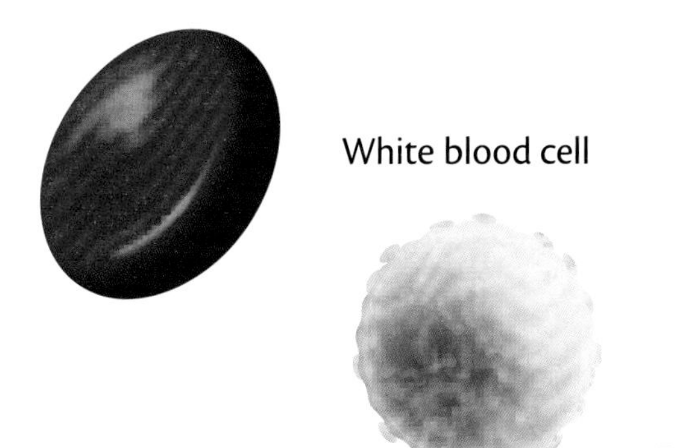

Plasma is the liquid that blood cells live in. Plasma carries food, salt, and water around your body. It also carries waste to your **kidneys**. Your kidneys clean the waste out of your blood. That waste goes out of you as a liquid called **urine**.

Do you remember what a cycle is? Blood travels through our bodies in a cycle when it does its jobs. It ends up at the place it started from and travels around again. Now let's learn how blood travels through your body.

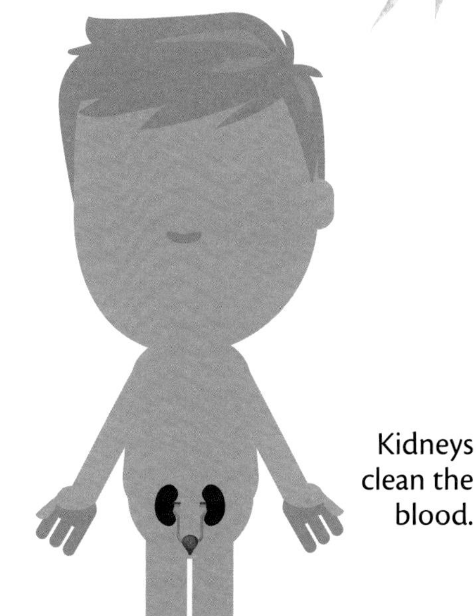

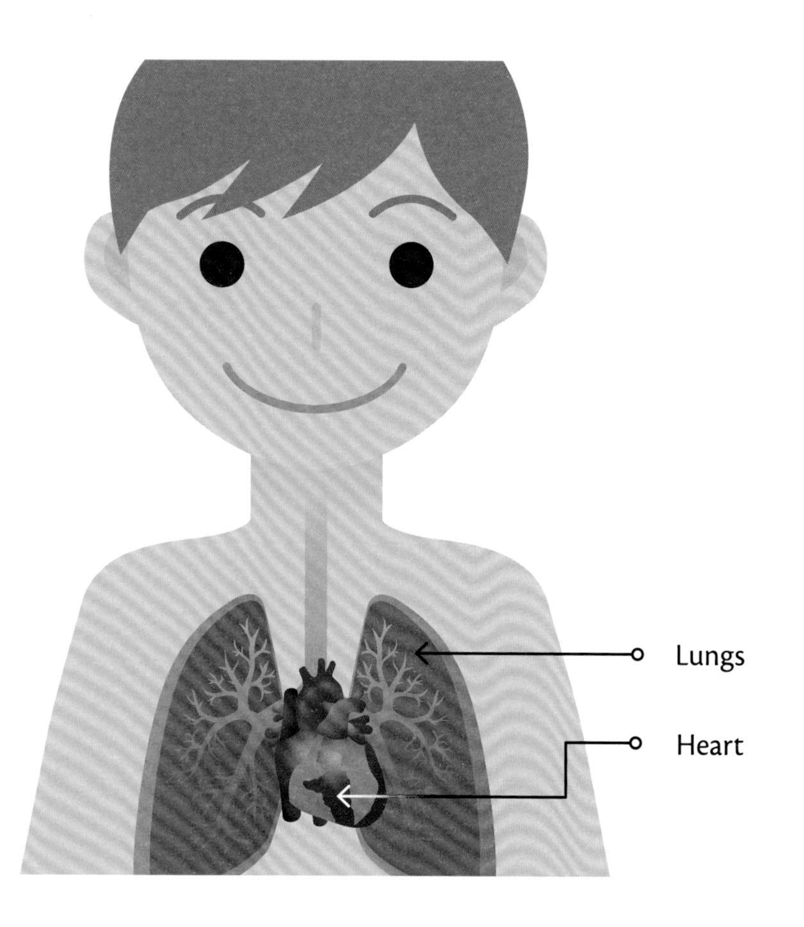

1. The **heart** is a muscle that makes blood move through our bodies. It is hollow and filled with blood. About once every second, it squeezes the blood out. Then it relaxes so it can fill with blood again. It does this automatically all day and all night long throughout your entire life!

2. From the heart, the blood travels through a tube to the **lungs**. The lungs are spongy sacks where air is squeezed in and out when you breathe. They give **oxygen** to the blood.

3. Then the blood travels back to the heart. The heart gives a big squeeze to push the blood into tubes that will take it around the body.

4. The tubes branch into smaller and smaller tubes. At the smallest tubes, oxygen leaves the blood cells and goes into your body's cells.

5. **Carbon dioxide** jumps into the red blood cells at the same time the oxygen leaves.

6. The blood travels back to your heart. Your heart squeezes and pushes it to the lungs again. There, it lets go of the carbon dioxide and picks up more oxygen for another cycle around your body!

Time to do Activity 104 in the Activity Book!

If we walk in the light as He is in the light, . . . the blood of Jesus Christ His Son cleanses us from all sin. (1 John 1:7)

Your blood keeps you alive while you live on Earth. Jesus' precious blood can give you life forever in Heaven! This verse tells us His blood can clean away the bad things you have done.

Your Food Is Changed to Energy

You get hungry! Food is healthy and yummy, so you eat. Then you are not hungry anymore. Eating is something you do on purpose. But **digesting** is something your body does automatically.

Your body needs to get four things from its food:

1. **Protein** comes from meat, nuts, eggs, and things made from milk. Many vegetables can also give you protein.
2. **Fat** comes from oils found in plants (like nuts and olives) and from animal foods (like meat and things made from milk).
3. **Carbohydrates** are sugars and starches that mostly come from plants. Milk also has a special kind of carbohydrate in it.
4. **Vitamins** and **minerals** are things in food that our bodies need in small

Definition

Digesting is the way your body automatically does all the things it needs to to make food into something you can use.

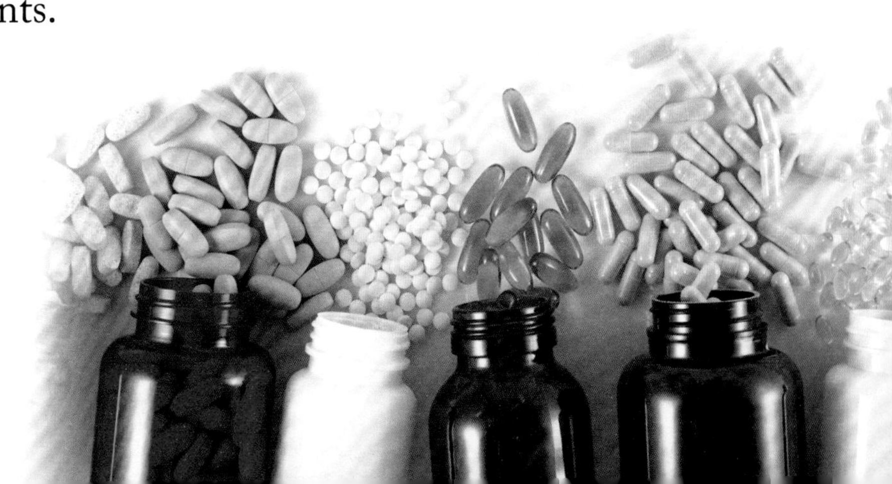

Sometimes we take vitamin pills in case we are not getting enough vitamins in our food.

amounts. Your body can turn protein, fat, and carbohydrates into energy. Vitamins and minerals do not change into energy. But each vitamin and mineral has its own important job to do in your body.

Most of your digestion happens because of **enzymes**.

Let's follow your food as it is digested:

Definition

Enzymes are proteins that speed up jobs in the body. Enzymes used for digestion help dissolve food and change it into what the body needs.

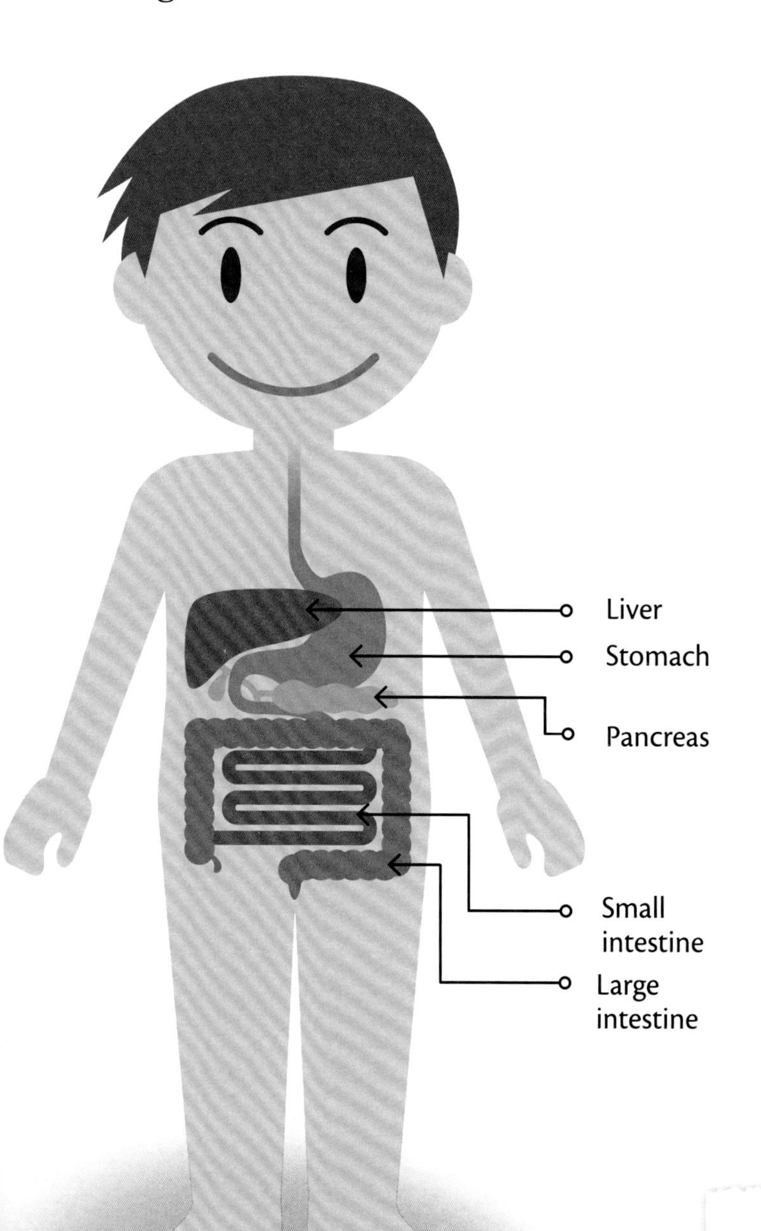

1. In your **mouth**, your food is chewed to make it smaller. Chewing helps enzymes do their job. The first enzymes that work on food come from your **saliva**. Saliva helps digest carbohydrates.
2. When you swallow, the food travels down a tube. The tube has muscles that automatically squeeze the food downward.
3. Your **stomach** opens to let the food in, and then it closes again. Then your stomach works on the food by squirting enzymes into it. It also squeezes to slosh the food around. Most stomach enzymes digest protein.

4. Next, the bottom of your stomach opens up to let the food into your **small intestine**. The small intestine is a long tube. Different enzymes squirt into the small intestine to digest fats, proteins, and carbohydrates. These enzymes come from the places they are made (like the **liver** and **pancreas**). Then the energy, vitamins, and minerals from the digested food travel into the blood through the sides of the small intestine. The blood takes them to the body's cells.

5. Next, the bottom of your small intestine opens up. Water and the parts of food your body can't use travel into your **large intestine**. The large intestine is a shorter, wider tube. Water goes out through the sides of your large intestine into your blood. The blood takes the water to the rest of your body. The solid waste left in the large intestine travels out of your body when you use the toilet.

Prayer

Lord, I am so glad I don't have to purposely make everything in my body work. So many of my body's jobs are automatic. But You are really the One making them work. Thank You for making me grow. Thank You for killing germs inside me that could make me sick. Thank You for my digestion and for the way my blood delivers good things to my cells. Amen.

Time to do Activity 105 in the Activity Book!

CHAPTER 36

Moving and Learning

Some parts of your body don't work automatically. They wait for your brain to tell them to do their jobs. When you wake up, you decide if you will get out of bed right then or lie there remembering the dream you had. But if your parents tell you to get up, you need to make your body obey right away. You are making your bones and muscles move on purpose.

Without bones and muscles, our bodies couldn't move. We couldn't walk, run, jump, ride a bike, turn our head, use our fingers, or chew a carrot. How could you play if God didn't give you muscles and bones? God wants you to play because playing helps you learn. God also wants you to use your muscles and bones to do work.

Whatever you do, work heartily, as for the Lord and not for men. (Colossians 3:23 ESV)

Let's learn about God's great invention of bones and muscles!

Bones

Bones are alive. Your bones are made of living cells. Bone cells connect themselves together with a lot of bridges growing out of their edges. The bridges are made of hard minerals. This is why bones are so hard and strong. There is space between each cell so your bones won't be too heavy.

God gave bones important jobs:

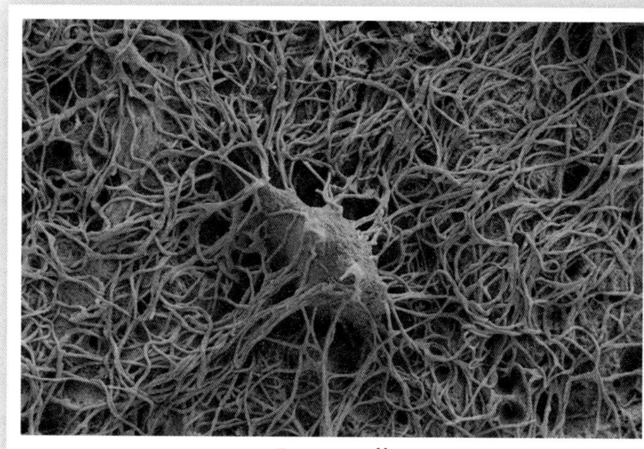

Bone cell

Bones hold you up so you don't have to lie on the ground like a worm.

Bones protect important parts of your body. Your rib bones are wrapped around your lungs and heart to protect them. Your skull bones cover your brain and keep it safe.

Bones store minerals. They give the minerals to your body when you need a little extra for certain jobs.

Blood cells are made in **bone marrow**. Marrow is a soft, spongy place in the middle of a bone.

Bones give your muscles something to attach to so you can move.

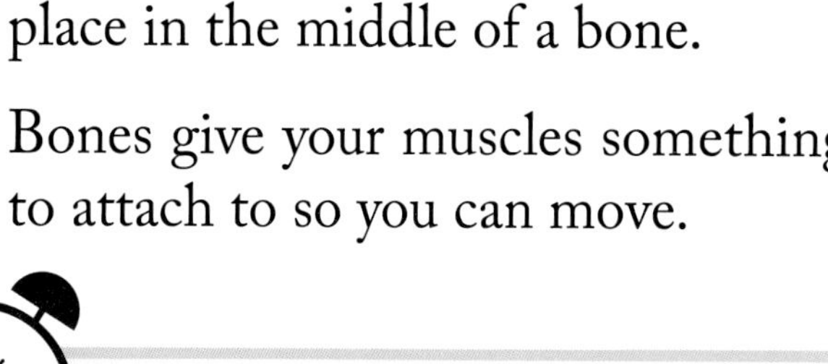

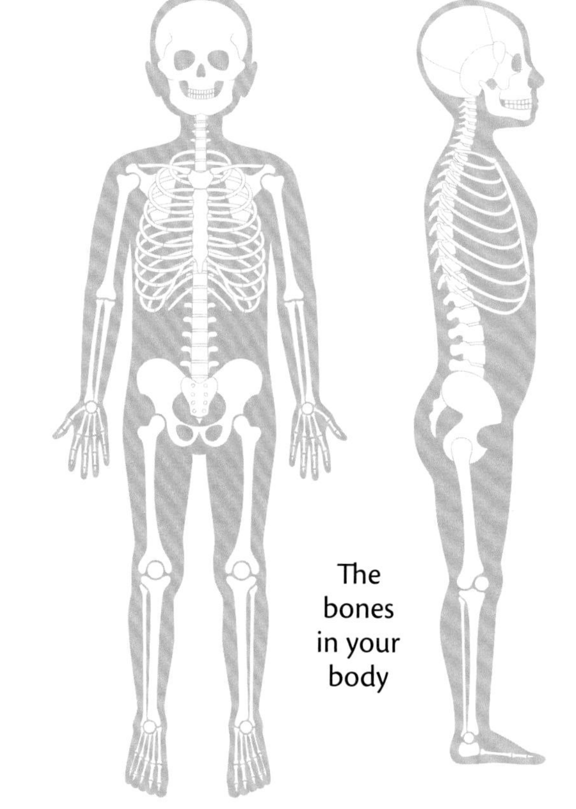

The bones in your body

Time to do Activity 106 in the Activity Book!

Muscles Move Your Bones

Muscles are made of long cells. Each muscle cell can **contract** and **relax.** When a muscle is used, its cells all contract or relax at the same time.

Definition

A muscle **contracts** by getting shorter and thicker.
When it **relaxes**, the muscle gets longer and thinner.

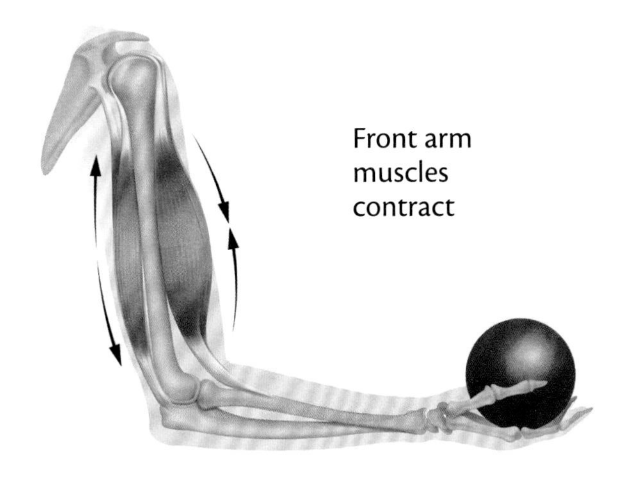

Let's learn what your body does when you eat a spoonful of ice cream.

1. Your **eyes** see the ice cream in your spoon and send a message through a **nerve** to your **brain**.
2. Your **brain** sends a message through **nerves** to tell your arm **muscles** what to do.
3. Your front arm **muscles** contract.
4. At the same time, your back arm **muscles** relax.

5. This makes your elbow bend. Your arm **bones** are brought together. The ice cream is brought to your **mouth**, and you eat it!
6. Your **mouth** and **nose** send messages about the ice cream through **nerves** to your **brain**.
7. Your **brain** helps you understand that you want another bite.
8. Your **brain** tells your back arm **muscles** to contract and tells your front arm **muscles** to relax. This sends your spoon back down to the ice cream in the bowl. This is repeated over and over again until your ice cream is gone.

Time to do Activity 107 in the Activity Book!

Learning

When you were born, you already knew how to use the muscles of your lips and tongue to drink milk. You didn't have to learn. God made you that way. But since then, you have had to learn how to use your muscles and bones to crawl, walk, and feed yourself. Maybe you are learning to throw a ball or play the piano. Each time you do a certain motion, you are getting better at doing it. Now you can eat without making a mess. You don't fall as much when you run. Your fingers work better when you color a picture. You are learning to use your muscles and bones better by **practicing**.

Definition

Practicing is doing something over and over again while trying to do it better each time.

Boy practicing writing

Doing something over and over again doesn't just make it easier. It actually changes your brain. The neurons you are using get faster at sending information. They also make pathways for themselves that make it easier to send information the next time. When you first learned to walk, you had to think about moving your feet while balancing. Now you can walk without even thinking about it! That's because your brain changed. Isn't it fun to learn to do new things with your body?

This messy baby is still learning how to feed himself.

You are also learning to *know* things with your brain. We usually learn the things we know by using our eyes and ears. Your parents might have sung the ABCs to you over and over when you were younger. You listened and could soon sing it yourself. Now that you are older, you are looking at the letters, learning how to read the words they make, and how to write them.

Learning to ride a bike

Learning to know things also takes practice. Your brain changes as you practice a memory verse, math facts, or reading. Practicing helps you remember things.

What you have learned and received and heard and seen in me—practice these things, and the God of peace will be with you. (Philippians 4:9 ESV)

The best reason to practice reading is so you can read God's Word.

God doesn't turn your brain off when you sleep. Your neurons are busy all night long. They take the things you learned that day and turn them into things you will remember. While you are sleeping, the neurons that have new information make their electricity go backwards. This makes these neurons more tightly connected to each other the next day. When you wake up in the morning, their electricity will go forward again, and they will work better. Sleep helps you learn!

Another way to learn something is to practice a little each day. You will probably remember a song your whole life if you sing it a few times a day for a week. You won't remember it as well if you sing it twenty times in one hour and never sing it again.

Prayer

Dear Lord, You are so amazing! We have learned about many of the wonderful ways You make everything in this world work perfectly. Even the way we learn these things is amazing. I praise You for making me so wonderfully. I'm glad I can move and think! Thank You for Your special care of me and my body. Amen.

Time to do Activity 108 in the Activity Book!

NOTES

1. https://www.universetoday.com/13562/how-long-does-it-take-to-get-to-the-moon
2. https://www.grc.nasa.gov/www/k-12/Numbers/Math/Mathematical_Thinking/seeing_the_earth_moon.htm
3. https://www.healthline.com/health/how-far-can-the-human-eye-see
4. https://www.healthline.com/health/how-far-can-the-human-eye-see
5. https://www.healthline.com/health/how-far-can-the-human-eye-see
6. https://www.healthline.com/health/how-far-can-the-human-eye-see
7. https://discoveryeye.org/20-facts-about-the-amazing-eye
8. https://answersingenesis.org/geology/radiometric-dating/radiohalos-and-diamonds
9. Morris, John D. "Lots of Grand Canyons." *Acts and Facts*, vol. 43, no. 6, Jun. 2014, p. 18.
10. https://en.wikipedia.org/wiki/Wind_wave
11. https://answersingenesis.org/kids/fish/ocean-box
12. https://www.nature.com/scitable/knowledge/library/water-uptake-and-transport-in-vascular-plants-103016037/
13. https://www.npr.org/sections/thetwo-way/2016/08/05/488891151/the-mystery-of-why-sunflowers-turn-to-follow-the-sun-solved
14. https://phys.org/news/2019-08-trees-chemical-beetles.html
15. Tomkins, Jeffrey P. "Fish as Smart as Apes?" *Acts & Facts*, vol. 45, no. 12, Dec. 2016, p. 15.
16. https://www.bbc.com/future/article/20191031-the-truth-behind-why-zebras-have-stripes

LIST OF IMAGES

Image Source Key:
iStock.com: i
unsplash.com: u
Wikimedia Commons: w
Image by author: a

---

11. Cat — u
12. Monkeys see the same colors you do. — u
13. Birds see more colors than you do. — u
14. Boy eating apple — i
15. Worm — i
16. Brothers — i

UNIT 2

1. Earth from space — i

CHAPTER 5

1. As they travel around the earth, man-made satellites gather information about the earth and space or help direct phone and television signals. — i
2. Earth — i
3. Boy looking at ocean — u
4. Whole Earth — i
5. Earth interior — i
6. 2014 Bárdarbunga volcanic eruption, Iceland — i
7. Some volcanoes have lava that flows instead of exploding. — u
8. No gravity — i
9. Astronaut floating in space where there is not enough gravity to pull him to Earth — u
10. Kids bordering Earth illustration — i

CHAPTER 6

1. Barren landscape — i
2. Crust illustration — i
3. Obsidian — i
4. Granite — i
5. Sandstone — i
6. Limestone — i
7. Marble — i
8. Slate — i
9. Silver and gold jewelry — i
10. Silver and gold coins — i
11. Copper Wire — i
12. Metal can be pounded and stretched — i
13. Cars, trains, and buildings made of metal — i
14. Steel cables on a bridge — i
15. Gems — i
16. Uncut diamond — i
17. Cut diamond — i
18. Uncut ruby — i
19. Cut ruby — i
20. God protects Noah's family while the flood changes the earth. — i
21. Mountains that were pushed up — u
22. Bent rock layers — i
23. Grand Canyon — u
24. Layered mountains — u
25. Rocks in sandstone — a
26. Volcanic mountain — u
27. Dinosaur fossil — u
28. Petrified tree — i

CHAPTER 7

1. Boy in ocean — u
2. Three views of Earth illustration — i
3. Land and ocean crust diagram — i
4. Dangerous wave caused by an earthquake — i
5. Curling wave — i
6. Rocks covered in oysters — i
7. Ocean plants — i
8. Heated water coming up from ocean floor — i
9. Dog — u
10. Water's three costumes — i
11. Water cycle diagram — i
12. A girl waters the dirt so her plant's roots can drink. — i
13. Boy ice fishing — i

CHAPTER 8

1. Four seasons — i
2. Earth's movement diagrams — i
3. Earth's tilt diagram — i
4. Illustrations of seasons — i
5. Gray whales swim about 12,000 miles each year. — i
6. Arctic terns fly 44,000 miles (71,000 km) each year. — a
7. Monarch butterflies fly about 4,800 miles (7,800 km) each year. — i
8. Bears — i
9. Bats — i
10. Groundhogs — i
11. Hedgehogs — u
12. The wood frog has special blood for living through frozen winters. — i
13. Bees — u
14. Penguin chicks — i
15. Rattlesnakes — u
16. Girl in snow — u

UNIT 3

1. Moon — u

CHAPTER 9

1. Flying kites — i
2. Dirty bowl — i
3. Eagle — u
4. Air is heated inside a balloon to make it rise. — i
5. Wing diagram — i

CHAPTER 10

1. Trees at sunset — u
2. God made five layers of sky. — i
3. Northern lights — u
4. Rain falling from clouds. — u
5. Sledding — u
6. Snowflakes come in many beautiful designs. — u
7. Girl in wind — i
8. If you see a tornado, go inside to a basement or inner room. — u

CHAPTER 11

1. Moon and cactus — u
2. Moon at beach — u
3. Illustration of Sun and Earth — i
4. Tide diagram — i
5. Girl looking in tide pool — i
6. During a total solar eclipse, the moon completely covers the sun. — i
7. Family observing stars — i

CHAPTER 12

UNIT 4

CHAPTER 13

CHAPTER 14

CHAPTER 15

CHAPTER 16

UNIT 5

CHAPTER 17

CHAPTER 18

CHAPTER 19

CHAPTER 20

UNIT 6

CHAPTER 21

7. Striped bass — i
8. Belly of cownose ray — i
9. Shark — i
10. A baby shark will hatch inside this shark's purse. — i
11. Sponge — i
12. Coral — i
13. Jellyfish — i
14. Anemone — i
15. A living Scotch bonnet mollusk with its shell on its back — i
16. Girl with seashells — i

CHAPTER 22

1. Boy looking at frog — i
2. Frog with eggs — i
3. Frogs can get oxygen through the skin of their mouths. — i
4. Frog development — i
5. Male frog calling a female — i
6. Frogs have webbed feet. — i
7. Toad catching an insect with its sticky tongue — i
8. Frog using eyeballs to swallow — i
9. Poisonous salamander — i
10. Nonpoisonous salamander — i
11. Salamander — i

CHAPTER 23

1. Horned toad warming itself — i
2. Crocodile scales — i
3. A snake sheds its skin. — i
4. Sea turtle hatching — i
5. Snake eating a mouse — i
6. Rattlesnake tongue and pits — i
7. Chameleons have long tongues for catching insects. — i
8. Sea turtle — i
9. Turtle bones — i
10. Tortoise — i
11. Alligator — i
12. Crocodile — i
13. Baby crocodile on adult's head — i

CHAPTER 24

1. Dinosaurs — i
2. Dinosaurs babies were small. — i
3. Fossilized footprint — i
4. Fresco, St. George and the Dragon, Monastery Church — i
5. Argentinosauruses — i
6. Torosaurus — i
7. Tyrannosaurus rex — i
8. Young and old stegosauruses — i
9. Pterodactyl — i
10. Plesiosaur — i
11. Giant ichthyosaurs probably looked like porpoises. — i
12. Dinosaur drawing — Daniel Sechrist

UNIT 7

1. God cares for sparrows and for you! — i

CHAPTER 25

1. Children chasing seagulls — i
2. Feather — i
3. Baltimore oriole taking off — i
4. Preening parrot — w
5. Bridge with support structures like bird bones — i
6. Blackbird eating a blueberry — i
7. Air travels into a bird's lungs (pink) and air sacs (blue). — a
8. Bird digestion — i
9. Northern harrier hawk — u

CHAPTER 26

1. A king penguin checks its egg and keeps it warm on top of its feet. — i
2. A cardinal's beak is a seed cracker. — i
3. An eagle's beak is a meat tearer. — i
4. A hummingbird's beak is a flower probe. — i
5. A parrot's beak is a nut cracker. — i
6. A snakebird's beak is a fish spear. — i
7. A pelican's beak is a fish net. — i
8. A duck's bill is a water strainer. — i
9. A woodpecker's beak is a wood cutter. — i
10. Bird — i
11. Duck feet — i
12. Feet of bird of prey — i
13. Ruffed grouse in snow — i
14. Ostrich feet — i
15. Father robin taking a turn on the nest — i
16. Ostrich egg and chicken egg — i
17. Robin's nest — i

CHAPTER 27

1. Barnacle geese — i
2. Storks — i
3. Black swifts — i
4. Turtledoves — i
5. Pelican — i
6. Blue tit at its nest — i
7. Male cardinal feeding his babies — i

CHAPTER 28

1. A dipper has caught a good meal in the stream. — i
2. Hummingbird — i
3. Hummingbird with white tongue sticking out — i
4. Hummingbird on nest made of lichen and spiderweb — i
5. Ruby-throated hummingbird — i
6. Honeyguide — i
7. Worker bees tend to baby bees and honey in wax rooms of the hive. — i

UNIT 8

1. Boy with bunnies — i

CHAPTER 29

1. Bear nursing her cubs — i
2. Lioness feeding her cubs — i
3. Camel nursing — i
4. Deer in winter — i
5. Whiskers help a mammal feel its way in the dark. — i
6. Porcupine with its quills raised — i
7. Fighting wolves — i
8. Bighorn sheep fighting — i
9. Baby elephant — i
10. Newborn mice — i

CHAPTER 30

CHAPTER 31

CHAPTER 32

UNIT 9

CHAPTER 33

CHAPTER 34

CHAPTER 35

CHAPTER 36